USA Math Competiti

# AN EXPLORATION OF CHALLENGING AMC 10 PROBLEMS

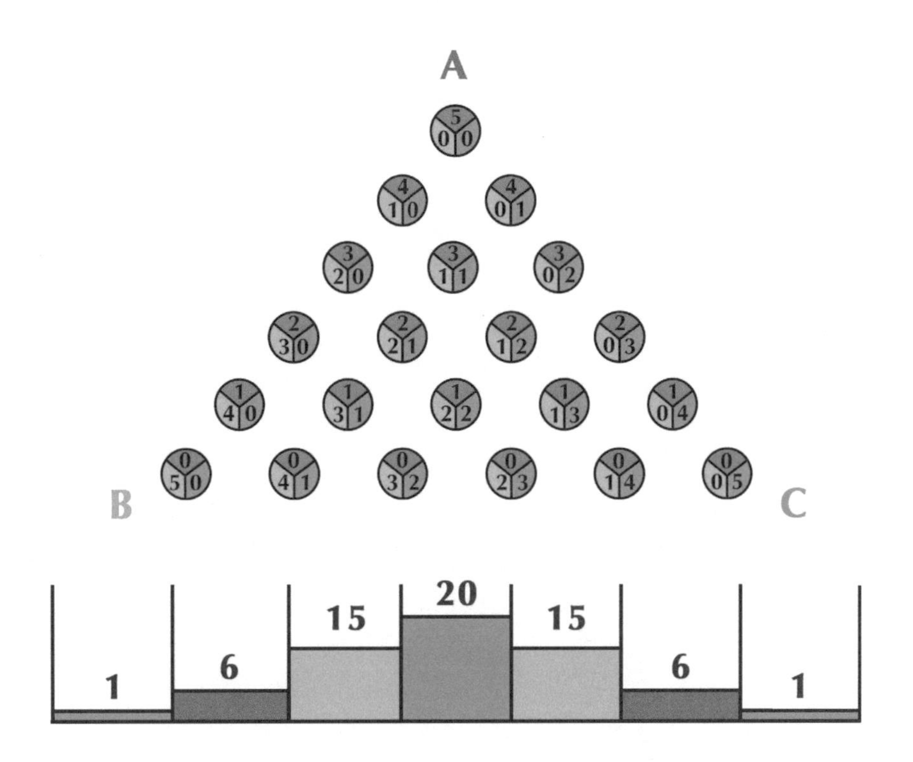


SIKUN "KEVIN" LAN
QIAO "TIGER" ZHANG

Printed in the United States of America

ISBN 979-8-9884562-0-9

# CONTENTS

# FOREWORD

Shortly after I launched *The Art of Problem Solving* website in 2003, I received an email from a former classmate of mine, saying, "I certainly wish your website and materials had existed when I was in high school. I went through junior high and high school without ever missing a question on a test, and then took [Math] 103 and 104 at Princeton, which was one of the most unpleasant and bewildering experiences of my life and poisoned me on math for years."

In this brief email, my classmate described why books like the one you're reading now are so important. While in college, I saw many students suffer through experiences like my classmate's. These students had aced math all through middle and high school, thinking, "I'm going to be a scientist. I'm going to be an engineer. I'm going to build amazing things." Then they enrolled in their first college math class or their first physics class and hit a wall they never saw coming.

A few students, like myself, were fortunate enough to have learned how to scale walls like these before we went to college. Many of us learned through math competitions. It wasn't the competitive nature of these events that was critical to our college preparation. It was the problems. Deep, rich, challenging problems that were unlike our school homework, unlike our school tests, and sometimes unlike any problems we had ever seen before.

Unlike those found at most high schools, test problems at top-tier colleges are, on the surface, nothing like the homework. Students can't simply memorize facts to make their way successfully through these exams. But they can, with training and creativity, tackle the test problems using the same fundamental tools they have used to complete their homework. Most students don't have any experience with this sort of problem solving prior to college. They've never been trained to develop a deep understanding of how their tools work, which would allow them to apply these tools to novel situations. *An Exploration of Challenging AMC* 10 *Problems* offers this training, providing the kind of insights and problem-solving heuristics that students need for later success in solving difficult problems, and not just in mathematics! Through *An Exploration of Challenging AMC* 10 *Problems*, Qiao "Tiger" Zhang and his teacher, Dr. Sikun "Kevin" Lan, share some of the strategies and critical thought processes that produced these strategies that have led to Tiger's great success in math competitions and his winning a Spirit of Ramanujan Fellowship in 2020. I look forward not only to Tiger's mathematical explorations in coming years, but also to his continued efforts in sharing his knowledge with those who hope to follow in his footsteps.

*Richard Rusczyk*
Founder and Chief Executive Officer
Art of Problem Solving Inc.
March 20, 2022

# INTRODUCTION

Welcome! This book was written for students who have prepared to take the AMC 10 and are hoping to qualify for the USAJMO, as well as their teachers and parents.

Chapters 1 to 4 address the topics of Geometry, Algebra, Counting and Probability, and Number Theory, respectively, and each chapter features different problem-solving techniques. The chapters can be read in any order. Each chapter begins with a conversation between the authors, Kevin and Tiger, about the title subject and the methods they used to help master the subject. This is followed by four to seven example problems. Each example includes a discussion of the problem and/or the techniques the authors used to find a solution, followed by one or two solutions. We also include some cartoons just for fun! Some examples have closing comments after the solutions are explained.

Each example is followed by about ten exercises, with a similar spirit to the example problem. Exercises span a wide range of difficulties. Full solutions to these exercises are available in the online version of the book (https://learnusamath.com). We recommend working on each exercise for at least 10-20 minutes before you read the solution.

Questions, new solutions, or any other topics you would like to discuss with Kevin or Tiger can be emailed to kevinlanteach@gmail.com.

# ABOUT THE AUTHORS

**Sikun "Kevin" Lan** was born in Shanghai, China and became interested in mathematics in elementary school. While attending The No. 2 High School Attached to East China Normal University, Kevin participated in dozens of math competitions at the district, city, national, and international levels and received many awards. After scoring in the top ten on Shanghai's College Entrance Exams STEM Division, he entered the Department of Mathematics at Peking University, where he earned a double Bachelor of Science degree in Mathematics and Physics. This was followed by a Master's degree and Ph.D. in applied mathematics that he received from UCLA.

Kevin then began a lengthy career as an IT specialist in banking, insurance, and digital marketing. During a 14-year stint as a software developer at Citibank, he participated in the development of global ATM machine software and online banking software, as well as the content management of the corporate and regional business websites of Citigroup (the parent company of Citibank).

In 2011, Kevin began coaching MathCounts individual competitors and the competition team at Gaspar De Portola Middle School, and shortly afterwards began his own online education business. Since then, he has tutored and coached many students participating in elementary, middle, and high school level math competitions, including the AMC 8/10/12 and AIME, as well as the SAT/ACT math and AP Calculus exams.

In 2017, Kevin took a position as math educator at the Geffen Academy, an independent secondary school associated with UCLA. In 2020, after earning a California Single Subject teaching credential, Kevin became a math teacher at Thousand Oaks High School in the Conejo Valley Unified School District. In 2022, Kevin continued teaching math at Calabasas High School in the Las Virgenes Unified School District.

**Qiao "Tiger" Zhang**, 15, is a 10th grader living in Southern California. Tiger grows up in a supportive family, although neither of his parents have any STEM-related background. He fell in love with math at a young age. In his childhood years, his favorite toys were logic games, and he enjoyed spending hours mastering the games and creating his own levels. In 2nd grade, Tiger joined the UCLA Math Circle (now UCLA Olga Radko Endowed Math Circle), his first and only math community for many years.

Growing up, Tiger loves reading books and solving challenging math problems. Inspired by his reading and thinking, he started writing a book he calls *Cool Math* in 2nd grade. It records special numbers and patterns, shortcuts to problems, and proofs of theorems and formulas he didn't know at the time.

Tiger's passion for math naturally prepares him for math competitions. Tiger took the AMC 10 for the first time in 6th grade and qualified for the AIME. He qualified for the USAJMO in 7th and 8th grade and the USAMO in 9th and 10th grade. In 9th grade, Tiger participated in MOP after getting a USAMO bronze medal and qualified for the USA TST. In 10th grade, Tiger received a gold medal for the USAMO. Tiger also achieved a perfect score in the AMC 10B as an 8th grader, the only perfect score in the world among 20,000 participants. Because of his curiosity, creativity, and pure passion for math, Tiger was selected twice as a recipient of the Spirit of Ramanujan Fellowship in 2020 and 2022. He dreams of becoming a person like Ramanujan who makes breakthroughs in math.

Besides math, Tiger loves music and is a passionate young pianist. He is a merit scholarship student at the Colburn Community School of Performing Arts. In 2022, Tiger performed as a soloist with the MTAC Glendale Concerto orchestra. He was also a semifinalist of the 2022 Music Center's Spotlight Award. A member of the Nth Trio, Tiger and his partners were Horszowski Piano Trio Prize winners in the 50th Fischoff Chamber Music Competition. The trio has played at various music events throughout the region. In his spare time, Tiger loves cycling, watching movies, and playing basketball, chess, and board games with his friends. He also dedicates much of his time to teaching students at the UCLA Math Circle. Tiger also likes to create math problems for various math classes and competitions. Some of the problems are used in this book as exercises.

# A NOTE TO TEACHERS

It can be difficult to coach students doing math at the AMC 10 level or above, and many of the problems are very challenging for teachers, including the teacher/author of this book. This is one reason we wrote a book that provides a way for students to engage in self-study.

However, the exercises contained in this book are often not enough for students who are new to these concepts/techniques to gain a thorough understanding of them. Thus, we encourage you to supply them with additional exercises, especially some that are easier than the examples included in the book, which can serve as stepping stones.

While this book contains many of the teaching philosophies and approaches that Kevin follows, they are by no means instructionally unique or applicable to every student. Any thoughts or experiences you would like to share with Kevin, as well as any questions, can be emailed to kevinlanteach@gmail.com.

# A BRIEF INTRODUCTION TO THE MAA MATHEMATICS COMPETITIONS SERIES

**THE MATHEMATICAL ASSOCIATION OF AMERICA** (MAA, https://maa.org) is the world's largest community of mathematicians, students, and enthusiasts that was established in 1915. Its American Mathematics Competitions are a series of examinations including the AMC 8, the AMC 10, the AMC 12, the AIME, the USAJMO, and the USAMO and curriculum materials that build problem-solving skills and mathematical knowledge in middle and high school students (https://maa.org/math-competitions). Following is a brief description of each examination.

**AMC 8** (American Mathematics Competition 8): An annual contest for 8th graders and younger. Each test contains 25 multiple-choice problems to be completed in 40 minutes.

**AMC 10** (American Mathematics Competition 10): An annual contest for 10th graders and younger. The two versions of the contest, the 10A and 10B, can be taken in the same year. Each test contains 25 multiple-choice problems to be completed in 75 minutes.

**AMC 12** (American Mathematics Competition 12): An annual contest for 12th graders and younger. The two versions of the contest, the 12A and 12B, can be taken in the same year. Each test contains 25 multiple-choice problems to be completed in 75 minutes.

**AIME** (American Invitational Mathematics Examination): Those who score well on the AMC 10 or AMC 12 are invited to participate in this annual contest. There are two versions: the AIME I and AIME II, which cannot be taken in the same year. Each test contains 15 problems, whose answers are integers between 0 and 999, to be completed in 3 hours.

**USAJMO** (United States of America Junior Mathematics Olympiad): Those whose scores on the AMC 10 and AIME of the year reach the cutoff for the year are invited to take this test. Each test contains 6 proof-based problems to be completed in 9 hours over the course of two days.

**USAMO** (United States of America Mathematics Olympiad): Those whose scores on AMC 12 and AIME of the year reach the cutoff for the year are invited to take this test. Each test contains 6 proof-based problems to be completed in 9 hours over the course of two days.

**IMO** (International Mathematics Olympiad): Those who score on the top of USAMO are invited to a summer training camp (the Mathematical Olympiad Program, or MOP). There, a team of six are selected to represent the USA to compete with other countries in the International Mathematical Olympiad (https://imo-official.org). Each test contains 6 proof-based problems to be completed in 9 hours over the course of two days.

# ACKNOWLEDGMENTS

**TIGER** I want to thank AoPS and Math Circle for creating a platform for math lovers that was my only math community for a long time. Thanks to my math teacher, Mr. Chris Tillman, whose passion for math has inspired me throughout my math career. Above all, thanks to my parents. Without their love and support, I wouldn't be where I am today.

**KEVIN** We want to thank Barry and Nadine Fox for coaching and editing work, Anna Wang for graphic design, and Michael Zeng for math proofreading.

*one*

# GEOMETRY

**TIGER** Geometry is my favorite subject.

**KEVIN** Cool, it's also my favorite subject. Why do you like it?

**TIGER** Because it's visual and beautiful, especially Olympiad geometry. Solving a geometry problem is like solving a puzzle; you use clues to build a diagram bit by bit until you see the whole picture. Also, I like the creativity needed to solve geometry problems.

**KEVIN** I first studied geometry, by myself, during the summer after finishing elementary school. In my final year of elementary school, I developed a tremendous interest in math inspired by my math teacher, Mr. Yu. I self-studied all through middle school and high school by reading math books during the summer.

When I studied geometry in school, we spent most of the time practicing conducting proofs, which was generally harder than doing problems that involved computing measurements of angle, length, area, and volume. In high school, I had opportunities to participate in a lot of math competitions—about 30 per year. I did so many geometry problems that I started to see geometry as a "gift," because whenever I saw a geometry problem in a competition, I had the confidence to solve it. That's the main reason I like geometry.

**TIGER** Another reason I like geometry is that I can make small observations about a diagram that begin to build up. Then, I can put everything together and solve the problem. It's a very rewarding experience.

**KEVIN** In recent years, about one-third of the problems in the AMC 10 in the 21-25 category have been geometry problems. Therefore, doing well in geometry problems is essential to being successful in the AMC 10. Back when I learned geometry, two-thirds of the problems we did were geometry proofs. But the AMC 10 has no proofs and schools in the U.S. rarely ask you to write proofs. Still, everything beyond AIME requires proofs. How did you find proof problems to work on?

**TIGER** I was exposed to proofs quite early. I attended the Art of Problem Solving classes, which had a graded proof problem at the end of each problem set. In these classes I got lots of valuable feedback that helped me improve my proof-writing skills. I also took Olympiad classes at the UCLA Math Circle, which introduced me to many techniques and ideas commonly used in proofs. I don't think proof problems are harder than computational problems because if you really solved the problem, you should also be able to write a proof. However, proof problems are somewhat different from computational problems because there are more proof-related techniques involved, such as contradiction.

Let's dive into our first example. This is one of the two problems I couldn't solve in the 2020 AMC 10B.

## Example 1.1
(AMC 10B 2020 Problem 21)

In square $ABCD$, points $E$ and $H$ lie on $\overline{AB}$ and $\overline{DA}$, respectively, so that $AE = AH$. Points $F$ and $G$ lie on $\overline{BC}$ and $\overline{CD}$, respectively, and points $I$ and $J$ lie on $\overline{EH}$ so that $\overline{FI} \perp \overline{EH}$ and $\overline{GJ} \perp \overline{EH}$. See the figure below. Triangle $AEH$, quadrilateral $BFIE$, quadrilateral $DHJG$, and pentagon $FCGJI$ each has area 1. What is $FI^2$?

(A) $\frac{7}{3}$ (B) $8 - 4\sqrt{2}$ (C) $1 + \sqrt{2}$ (D) $\frac{7}{4}\sqrt{2}$ (E) $2\sqrt{2}$

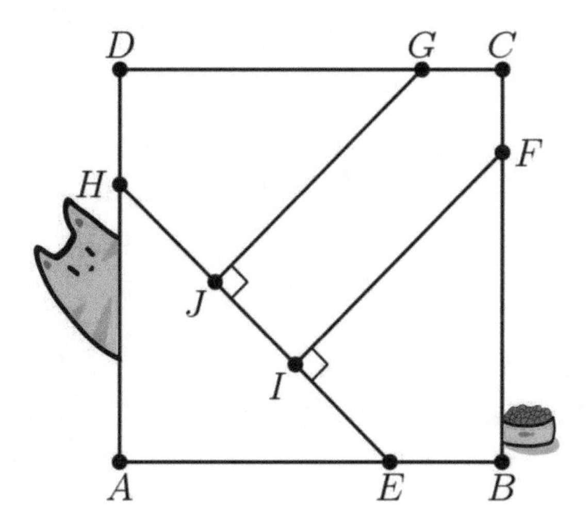


## Discussion

**KEVIN** 2020 was your second year participating in the AMC 10, and you were only in 7th grade. After your first attempt at the AMC 10 in 6th grade, you immediately qualified for the AIME. Many of my students have tried several times before qualifying. I often tell them to avoid making mistakes on easy problems, and I discourage them from trying problems 21-25 because you still get 1.5 points if you don't answer a problem. These problems often take a long time to solve and many students are unlikely to get them right.

Have you tried problems 21-25 in all of your AMC 10 tests so far?

**TIGER** Yes, I try most of the problems that look doable. I tried this problem for a few minutes in the contest and ended up with an ugly system of equations that I couldn't solve.

**KEVIN** This problem looks doable now, right? How would you approach this?

**TIGER** First, we need to learn as much as we can about this diagram. Since $AEH$, $BFIE$, $DHJG$, and $FCGJI$ add up to square $ABCD$, it must have area 4, so the square has side length 2.

Since $\Delta AEH$ has area 1, we have $AE = AH = \sqrt{2}$. Initially, I wanted to split $BFIE$ into $EFB$ and $EFI$, but that's quite tedious since it seems somewhat hard to relate $\overline{BF}$, $\overline{FI}$, and $\overline{IE}$. Instead, we can try to find other ways of getting the area of $BFIE$.

I thought of sliding $F$ on $\overline{BC}$ and $I$ on $\overline{HE}$ while maintaining $\overline{FI} \perp \overline{EH}$ until the area of $BFIE$ is exactly 1. As we move $\overline{FI}$, we see that the lines $IE$ and $FB$ do not change. So, we can take advantage of that by looking at their intersection, which is also constant. Suppose the intersection point is $K$, we find the area $[EBK] = [FIK] - [BEIF]$ doesn't depend on $F$ or $I$.

## Tiger's Solution

Because triangle $AEH$, quadrilateral $BFIE$, quadrilateral $DHJG$, and pentagon $CFIJG$ each has area 1, the square $ABCD$ must have area 4 and its side length is 2.

Triangle $AEH$ has area 1 and $AE = AH$, so $AE = AH = \sqrt{2}$, $EB = 2 - \sqrt{2}$.

Let $HE$ and $CB$ intersect at $K$. Both $\Delta FIK$ and $\Delta EBK$ are 45-45-90 triangles, so $\left[FIK\right] = \frac{1}{2}FI^2$ and $\left[EBK\right] = \frac{1}{2}\left(2 - \sqrt{2}\right)^2 = 3 - 2\sqrt{2}$.

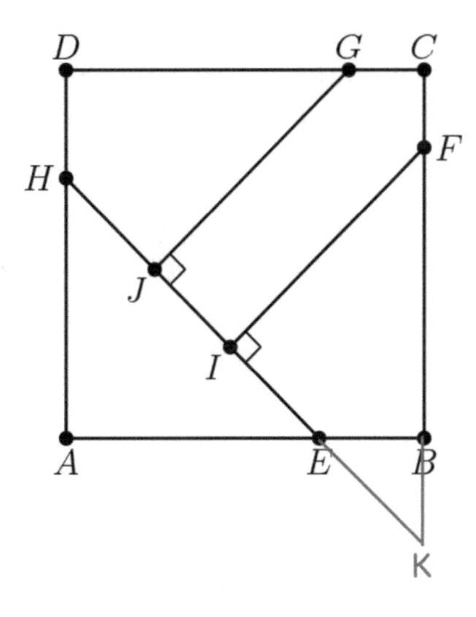


Since $[FIK] = [BFIE] + [EBK]$, $\frac{1}{2}FI^2 = 1 + 3 - 2\sqrt{2}$, so $FI^2 =$ (**B**) $8 - 4\sqrt{2}$.

**TIGER** The main idea of this problem is to look for nicer ways of representing geometric quantities. Extending lines and looking at constant areas made the solution a lot cleaner. In general, constructions in geometry can be tricky, but you should be ready for them when doing geometry problems. If a construction makes the problem simpler or cleaner, it's probably useful.

**KEVIN** I took a slightly different approach. I used a technique I call "rectangularization" to calculate areas, then set up a quadratic equation. This approach consists of dropping altitudes and creating rectangles to use perpendicularity, right triangles, and similar triangles to get information about the diagram. It can act as a sort of "coordinate system," but it's often cleaner than coordinates.

Rectangularization can also be used in problems with equilateral triangles by extending lines to get many $60^\circ$ and $120^\circ$ angles.

## Kevin's Solution

Because $AEH$, $BFIE$, $DHJG$, and $CFIJG$ each has area 1, $ABCD$ must have area 4 and side length 2. Therefore, we have $AC = 2\sqrt{2}$.

Triangle $AEH$ has area 1 and $AE = AH$, so $AE = AH = \sqrt{2}$ and $EH = 2$.

Let $\overline{AC}$ intersect $\overline{EH}$ at $L$, and $M$ be the foot of the perpendicular from $F$ to $\overline{AC}$.

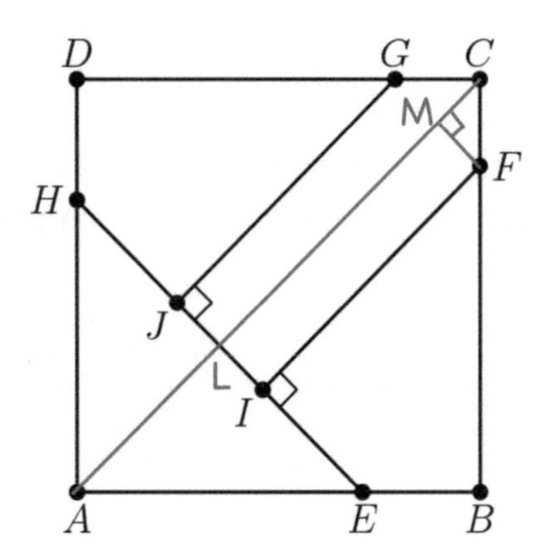


Let $x = FI$. Since $L$ is the midpoint of $\overline{HE}$, we have $AL = \frac{1}{2}HE = 1$,
Set $LM = FI = x$ to get $MF = MC = AC - AL - LM = 2\sqrt{2} - 1 - x$.

By symmetry, $\frac{1}{2} = \frac{1}{2}[FCGJI] = [FCLI] = \frac{1}{2}(2\sqrt{2} - 1 + x)(2\sqrt{2} - 1 - x)$.
By difference of squares, we have $(2\sqrt{2} - 1)^2 - x^2 = 1$, which means $x^2 =$ (**B**) $8 - 4\sqrt{2}$.

## Exercises: Horizontal and Vertical

**KEVIN** These exercises use similar tactics or techniques to the example above. They range from significantly easier to significantly harder than the example. You should try the problem until you solve it or have no ideas for a while. This should usually take at least 15 minutes.

**TIGER** As I study past competition problems, I like to create my own. Some of the exercises are from my problem collection.

## Exercise 1.1.1
(AMC 8 2015 Problem 25)

One-inch squares are cut from the corners of this 5-inch square. What is the area in square inches of the largest square that can be fitted into the remaining space?

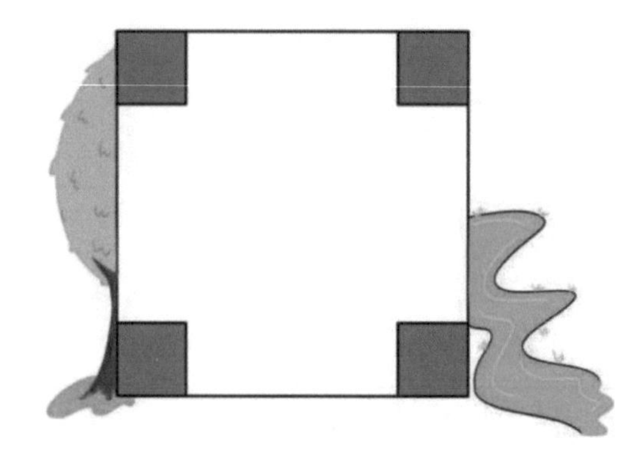

## Exercise 1.1.2
(Tiger)

In parallelogram $ABCD$, $AB = 13$, $BC = 14$, and the distance between $\overline{AD}$ and $\overline{BC}$ is 12. Point $E$ lies on $\overline{BD}$ such that $\angle ECB = 90°$. What is $CE$?

## Exercise 1.1.3
(Tiger)

$ABCD$ is a quadrilateral with $BA = BC$, $DA = DC$, $AC = 10$, and $BD = 18$. O is the circumcenter of $\Delta ABD$. If $\angle ACO = 90°$, what is CO?

### Exercise 1.1.4
(JMPSC Invitationals 2021 Problem 12, by Tiger)

Rectangle $ABCD$ is drawn such that $AB = 7$ and $BC = 4$. $BDEF$ is a square that contains vertex $C$ in its interior. Find $CE^2 + CF^2$.

### Exercise 1.1.5
(AMC 10B 2004 Problem 22)

A triangle with sides of 5, 12, and 13 has both an inscribed and a circumscribed circle. What is the distance between the centers of those circles?

### Exercise 1.1.6
(AMC 10A Spring 2021 Problem 21)

Let $ABCDEF$ be an equiangular hexagon. The lines $AB$, $CD$, and $EF$ determine a triangle with area $192\sqrt{3}$, and the lines $BC$, $DE$, and $FA$ determine a triangle with area $324\sqrt{3}$. The perimeter of hexagon $ABCDEF$ can be expressed as $m + n\sqrt{p}$, where $m$, $n$, and $p$ are positive integers and $p$ is not divisible by the square of any prime. What is $m + n + p$?

### Exercise 1.1.7
(HMMT November 2022 General Problem 7)

In circle $\omega$, two perpendicular chords intersect at a point $P$. The two chords have midpoints $M_1$ and $M_2$ respectively, such that $PM_1 = 15$ and $PM_2 = 20$. Line $M_1M_2$ intersects $\omega$ at points $A$ and $B$, with $M_1$ between $A$ and $M_2$. Compute the largest possible value of $BM_2 - AM_1$.

## Exercise 1.1.8
(AMC 12B 2014 Problem 21)

In the figure, $ABCD$ is a square of side length 1. The rectangles $JKHG$ and $EBCF$ are congruent. What is $BE$?

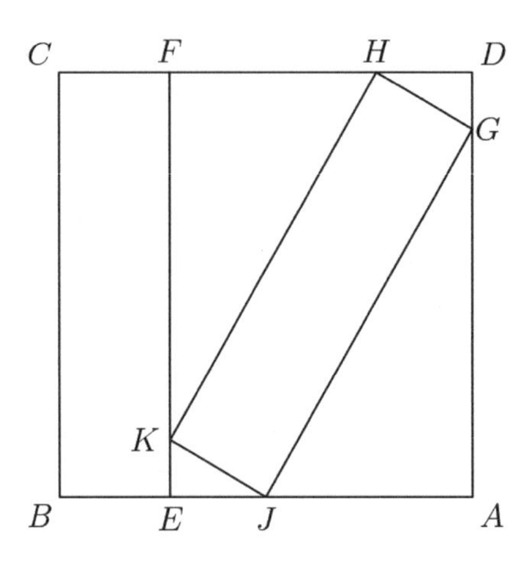


## Exercise 1.1.9
(Tiger)

Let $ABC$ be an equilateral triangle. Let $P$ and $Q$ be points such that $B$, $P$, $Q$, and $C$ are collinear in that order. Let $R$ lie on segment $CA$ and $S$ on segment $AB$. If $PQ = 5$, $PS = 3$, $QR = 4$, $\angle PQR = \angle QPS$, and $AR = BS$, find $AB$.

## Exercise 1.1.10
(AIME I 2015 Problem 7)

In the diagram below, $ABCD$ is a square. Point $E$ is the midpoint of $\overline{AD}$. Points $F$ and $G$ lie on $\overline{CE}$, and $H$ and $J$ lie on $\overline{AB}$ and $\overline{BC}$, respectively, so that $FGHJ$ is a square. Points $K$ and $L$ lie on $\overline{GH}$, and $M$ and $N$ lie on $\overline{AD}$

and $\overline{AB}$, respectively, so that $KLMN$ is a square. The area of $KLMN$ is 99. Find the area of $FGHJ$.

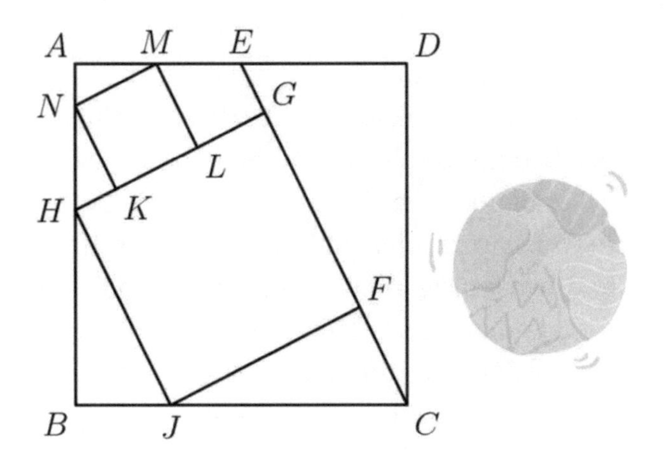


## Example 1.2
(AMC 10B 2017 Problem 24)

The vertices of an equilateral triangle lie on the graph of $xy = 1$ and $(1, 1)$ is the centroid of the triangle. What is the square of the area of the triangle?

(A) 48 (B) 60 (C) 108 (D) 120 (E) 169

## Discussion

**KEVIN** It took me five minutes to solve this problem. But in the AMC 10, you have only 75 minutes to complete 25 problems, an average of three minutes for each problem. So, most students can't afford to spend five minutes on Problem 24, even if they're able to solve it. How did you manage your time during the exam?

**TIGER** I was usually able to solve the first 15 problems in about 20 minutes.

**KEVIN** That's great. For students aiming for AIME qualification, I recommend that they spend the first 30 minutes solving the first 15 problems. If they do all of them correctly, they get 90 points. I also recommend that they leave problems 21-25 blank to get 7.5 points. This leaves them 45 minutes to work on problems 16-20. If they get three of them right and two wrong, they add another 18 points to their score, for a total of 115.5. This should qualify them for AIME. If they don't, it's often because they make mistakes in problems 1-15. How do you avoid making mistakes when doing easy problems?

**TIGER** It takes a lot of practice to reduce them. In a contest, actively check for mistakes while doing the problem. It's a lot easier to find a mistake before finishing the problem than to find it when you're going back and checking for mistakes. However, I think it's better to prioritize increasing the difficulty of the problems you can do rather than doing the easier problems with more accuracy since it's hard to practice being accurate.

**KEVIN** I advise my students to check problems 1-15 if they finish them before the 30-minute mark. Do you go back to check your completed problems during the exam? How do you decide whether to check your completed problems or to work on remaining problems?

**TIGER** In the last 10 minutes, I skim through the remaining problems and assess their difficulty, then decide whether to work on them or to start checking completed problems.

**KEVIN** I recommend most students to check their answers after completing problems 1-15 and before working on problems 16-25. If you're taking more than 5 minutes on any problem in the 1-15 range, skip it for now and come back to it when time is nearly over. It takes discipline to do so. Most students just go for the new problems and never have a chance to check easy problems.

How did you approach this problem?

**TIGER** One choice is to immediately try to solve it algebraically. However, the equilateral condition might be ugly to convert into algebra. Let's draw the graph of $xy = 1$ and try to roughly find where the equilateral triangle should be.

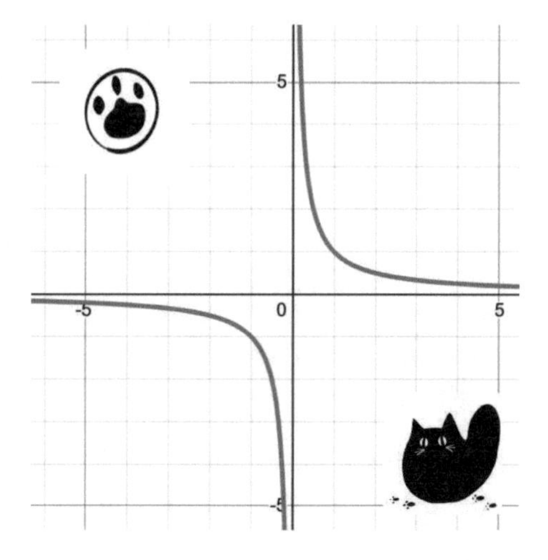

We know that the graph of $xy = 1$ is symmetric about the line $y = x$. Thus, it seems natural that the equilateral triangle would also be symmetric about $y = x$, which means one vertex of the triangle is at $(-1,-1)$. We can explain this visually: consider the circle centered at $(1,1)$ passing through the three vertices of the equilateral triangle. Then, the circle must intersect $xy = 1$ at least three times, once for each vertex. We see that the circle intersects $xy = 1$ at most four times. If there are four intersection points, they must form a quadrilateral symmetric about the line $y = x$, which we call $ABCD$, such that $A$ and $B$ are symmetric about $y = x$, and $C$ and $D$ are symmetric about $y = x$. If $ABC$ is an equilateral triangle, then $ABD$ is also an equilateral triangle, a contradiction since $C \neq D$. Thus, the circle must intersect $xy = 1$ exactly three times. This can only happen if the circle passes through $(-1,-1)$, so one of our vertices must be $(-1,-1)$.

Thus, we can calculate the distance from a vertex to the centroid, then find the length of each median. This gives the area of the triangle.

The main idea of this problem is to use a visual argument instead of an algebraic one. Note that we never found the other vertices of the triangle and it's not immediately obvious that they lie on $xy = 1$.

**KEVIN** I also used symmetry, but I did the rest algebraically. I used the fact that for a triangle with vertices $(x_1,y_1)$, $(x_2,y_2)$, and $(x_3,y_3)$, the coordinate of its centroid is $\left(\frac{1}{3}(x_1+x_2+x_3), \frac{1}{3}(y_1+y_2+y_3)\right)$.

## Tiger's Solution

Let the vertices of the equilateral triangle in quadrant I be $B$ and $C$, and let $A$ be the other vertex. By symmetry, we have $A = (-1, -1)$.

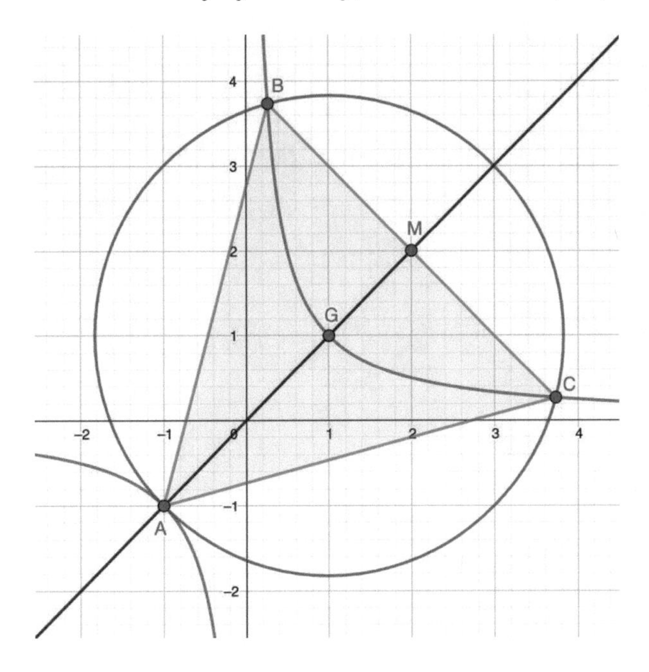


WWW.GEOGEBRA.ORG/CALCULATOR/ZJ3SSWB9

Let $G=(1, 1)$ be the centroid of $\Delta ABC$, and let $M$ be the midpoint of $\overline{BC}$. We have $AG = 2\sqrt{2}$. By symmetry, $\overline{AM}$ is an altitude of $\Delta ABC$. Since $G$ is the centroid, we have $AG = \frac{2}{3}AM$, so $AM = 3\sqrt{2}$. Since $\Delta ABM$ is a 30-60-90 triangle, we have $BM = \frac{AM}{\sqrt{3}} = \sqrt{6}$, so $BC = 2\sqrt{6}$. Therefore, we have $\left[ABC\right] = \frac{1}{2}BC \cdot AM = 6\sqrt{3}$, so $[ABC]^2 =$ (C)108.

## Kevin's Solution

Let the vertices of the equilateral triangle in quadrant I be $B$ and $C$, and let $A$ be the other vertex.

By symmetry, we can set the coordinates of $A$, $B$, and $C$ as $(-1,-1)$, $(x_1, \frac{1}{x_1})$, and $(\frac{1}{x_1}, x_1)$. Since the triangle's centroid is at $(1, 1)$, we have $\frac{1}{3}\left(x_1+\frac{1}{x_1}-1\right)=1$, or $x_1^2-4x_1+1=0$. The root with $x_1<\frac{1}{x_1}$ is $x_1=2-\sqrt{3}$.

Let $s$ be the side length of the triangle. Then, $s^2=AB^2=(x_1+1)^2+\left(\frac{1}{x_1}+1\right)^2=(3-\sqrt{3})^2+(3+\sqrt{3})^2=24$, so $[ABC]^2=\left(\frac{\sqrt{3}}{4}s^2\right)^2$ $=(6\sqrt{3})^2=$ (C)108.

## Exercises: Analytic Geometry and Symmetry

### Exercise 1.2.1
(AMC 8 2003 Problem 25)

In the figure, the area of square $WXYZ$ is 25 cm$^2$. The four smaller squares have sides 1 cm long, either parallel to or coinciding with the sides of the large square. In $\Delta ABC$, $AB=AC$, and when $\Delta ABC$ is folded over side $\overline{BC}$, point $A$ coincides with O, the center of square $WXYZ$. What is the area of $\Delta ABC$, in square centimeters?

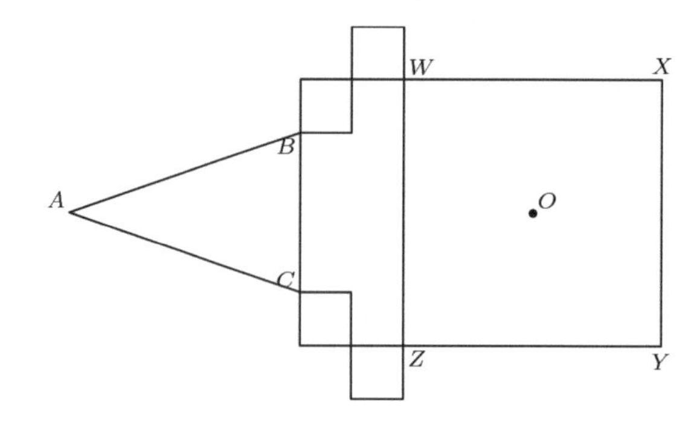

## Exercise 1.2.2
(Tiger)

$ABCD$ is a unit square with center O. $E$ and $F$ are on segments $\overline{AB}$ and $\overline{CD}$, respectively, such that the area of $AEFD$ is $\frac{1}{3}$ and $\overline{EF}$ bisects $\overline{AO}$. What is $EF$?

## Exercise 1.2.3
(AMC 10A Spring 2021 Problem 19)

The area of the region bounded by the graph of

$$x^2 + y^2 = 3|x - y| + 3|x + y|$$

is $m + n\pi$, where $m$ and $n$ are integers. What is $m + n$?

## Exercise 1.2.4
(AMC 10A 2013 Problem 18)

Let points $A = (0,0)$, $B = (1,2)$, $C = (3,3)$, and $D = (4,0)$. Quadrilateral $ABCD$ is cut into equal area pieces by a line passing through $A$. This line intersects $\overline{CD}$ at point $\left(\frac{p}{q}, \frac{r}{s}\right)$ where these fractions are in lowest terms. What is $p + q + r + s$?

## Exercise 1.2.5
(AMC 10B 2016 Problem 21)

What is the area of the region enclosed by the graph of the equation $x^2 + y^2 = |x| + |y|$?

## Exercise 1.2.6

(Tiger)

Let $x$ and $y$ be real numbers with $x^2 + y^2 - 11x - 8y + 44 = 0$ such that the value of $\frac{y}{x}$ is maximized. Find $11x + 8y$.

## Exercise 1.2.7

(AMC 10A Spring 2021 Problem 24)

The interior of a quadrilateral is bounded by the graphs of $(x + ay)^2 = 4a^2$ and $(ax - y)^2 = a^2$, where $a$ is a positive real number. What is the area of this region in terms of $a$, valid for all $a > 0$?

## Exercise 1.2.8

(AIME I 2006 Problem 10)

Eight circles of diameter 1 are packed in the first quadrant of the coordinate plane as shown. Let region $\mathcal{R}$ be the union of the eight circular regions. Line $l$, with slope 3, divides $\mathcal{R}$ into two regions of equal area. Line $l$'s equation can be expressed in the form $ax = by + c$, where $a$, $b$, and $c$ are positive integers whose greatest common divisor is 1. Find $a^2 + b^2 + c^2$.

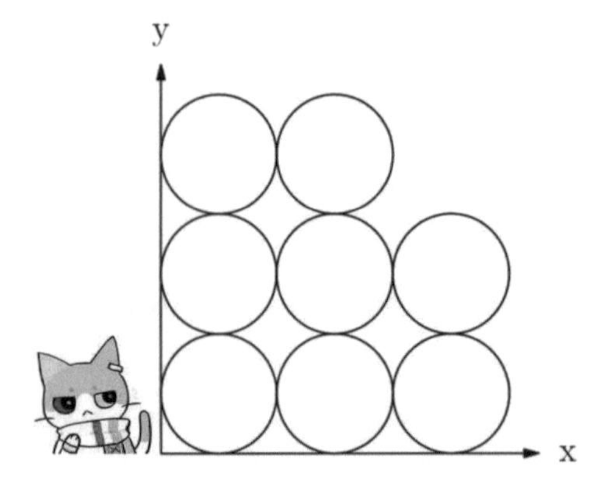

## Exercise 1.2.9
(USAJMO 2021 Problem 4 Modified)

Carina has three pins, labeled A, B, and C, respectively, located at the origin of the coordinate plane. In a *move*, Carina may move a pin to an adjacent lattice point at distance 1 away. Find, with proof, the least number of moves that Carina can make in order for triangle ABC to have area 5000.

## Exercise 1.2.10
(AIME II 2006 Problem 15)

Given that $x$, $y$, and $z$ are real numbers that satisfy

$$x = \sqrt{y^2 - \frac{1}{16}} + \sqrt{z^2 - \frac{1}{16}}$$

$$y = \sqrt{z^2 - \frac{1}{25}} + \sqrt{x^2 - \frac{1}{25}}$$

$$z = \sqrt{x^2 - \frac{1}{36}} + \sqrt{y^2 - \frac{1}{36}}$$

and that $x + y + z = \frac{m}{\sqrt{n}}$, where $m$ and $n$ are relatively prime positive integers and $n$ is not divisible by the square of any prime, find $m + n$.

## Example 1.3
(AMC 10A 2015 Problem 21 & AMC 12A 2015 Problem 16)

Tetrahedron $ABCD$ has $AB = 5$, $AC = 3$, $BC = 4$, $BD = 4$, $AD = 3$, and $CD = \frac{12}{5}\sqrt{2}$. What is the volume of the tetrahedron?

(A) $3\sqrt{2}$ (B) $2\sqrt{5}$ (C) $\frac{24}{5}$ (D) $3\sqrt{3}$ (E) $\frac{24}{5}\sqrt{2}$

## Discussion

**KEVIN** This is a 3-D geometry problem. From elementary school to my qualifying exams for Ph.D. candidacy, I got an F on only one math test—it was a 3-D geometry test in middle school. Do you think 3-D problems are harder than 2-D problems? How would you usually approach this type of problem?

**TIGER** We usually analyze them by cutting planes through shapes and looking at the cross-sections they make. This turns a 3-D problem into 2-D problems that we can analyze using plane geometry techniques. 3-D problems are usually less complex than 2-D problems, which often have elaborate steps involving cyclic quadrilaterals and similar triangles. Thus, I usually attempt 3-D geometry problems before 2-D geometry problems on the AIME.

**KEVIN** In 3-D geometry problem, the first step is to draw a good diagram. I often sketch a diagram to learn about the configuration, then draw a second diagram that I use to solve the problem.

**TIGER** Here, notice that since $AC = AD$ and $BC = BD$, the tetrahedron is symmetric about the plane containing $A$ and $B$ perpendicular to $\overline{CD}$. So, we should draw the tetrahedron such that $\overline{AB}$ is vertical and $C$ and $D$ are on opposite sides of $\overline{AB}$.

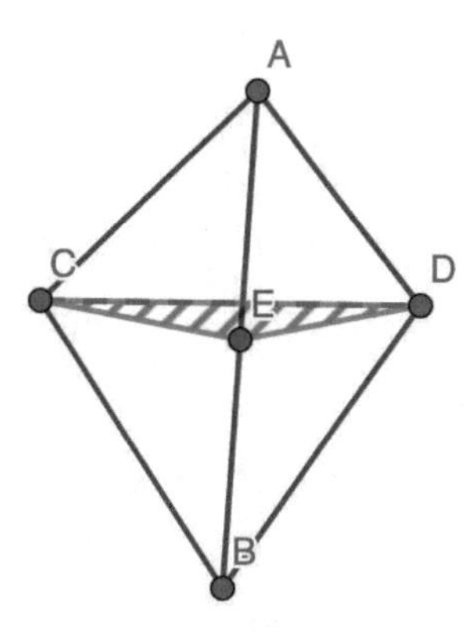


WWW.GEOGEBRA.ORG/CALCULATOR/V44DHNFT

**TIGER** This makes it more obvious that we can split the tetrahedron into two pieces by cutting a plane containing $\overline{CD}$ perpendicular to $\overline{AB}$. Let the plane intersect $\overline{AB}$ at $E$. Then, $\overline{AE}$ and $\overline{BE}$ are altitudes of these two tetrahedra. We can find the areas of both of these tetrahedra.

**KEVIN** I cut the tetrahedron in a different way; vertically instead of horizontally. Let $F$ be the midpoint of $\overline{CD}$.

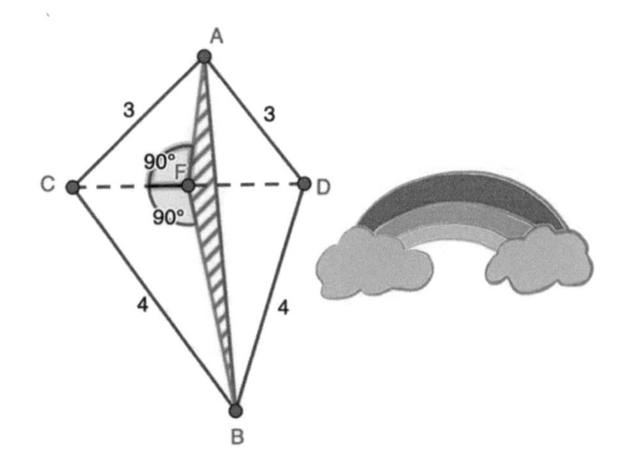


WWW.GEOGEBRA.ORG/CALCULATOR/F78TJJCC

**KEVIN** Since both $\Delta ACD$ and $\Delta BCD$ are isosceles triangles, we have $\overline{AF} \perp \overline{CD}$ and $\overline{BF} \perp \overline{CD}$. Therefore, plane $ABF$ cuts the tetrahedron into two smaller tetrahedra that share the base $\Delta ABF$ and whose altitudes are $\overline{CF}$ and $\overline{DF}$, respectively. We can calculate the volume of both tetrahedra.

**TIGER** The motivation for constructing plane $CDE$ is to make the diagram as nice as possible. It's hard to find the area just by dropping a perpendicular from a point, like $C$, to the plane containing the other three points, like $A$, $B$, and $D$.

## Tiger's Solution

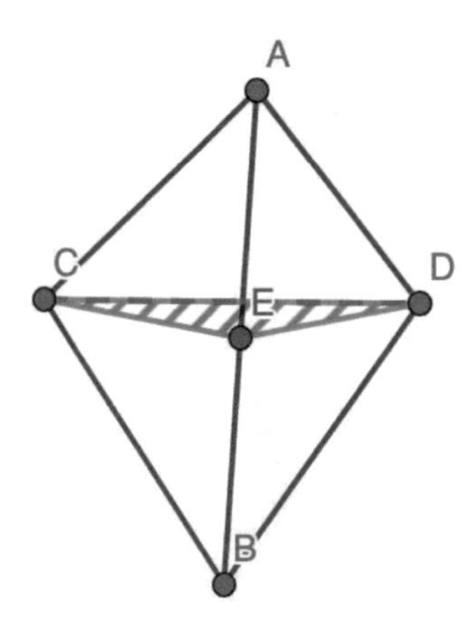


WWW.GEOGEBRA.ORG/CALCULATOR/F78TJJCC

Since $ABC$ and $ABD$ are congruent by SSS, there exists a point $E$ on $\overline{AB}$ such that $\overline{CE} \perp \overline{AB}$ and $\overline{DE} \perp \overline{AB}$. Thus, $\overline{AB}$ is perpendicular to plane containing $\Delta CDE$.

Since $\Delta ABC$ is right, its area is $\left[ABC\right] = \frac{1}{2} \cdot AC \cdot BC = 6$. We also have $\left[ABC\right] = \frac{1}{2} \cdot AB \cdot CE = \frac{5}{2} CE$, so $CE = \frac{12}{5}$. Similarly, $DE = \frac{12}{5}$.

Notice that since $CD\sqrt{2} = CE$, $\Delta CDE$ is a 45-45-90 triangle. Hence, its area is $\frac{1}{2}\left(\frac{12}{5}\right)^2 = \frac{72}{25}$.

Thus, the volume of tetrahedron $ABCD$ is sum of the volumes of the pyramids $ACDE$ and $BCDE$, which is $\frac{1}{3} \cdot \frac{72}{25} \cdot 5 = \text{(C)}\frac{24}{5}$.

## Kevin's Solution

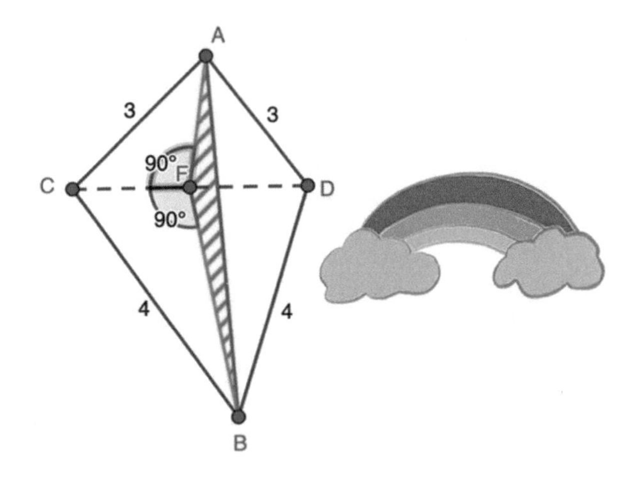


WWW.GEOGEBRA.ORG/CALCULATOR/F78TJJCC

Let $F$ be the midpoint of $\overline{CD}$. Since $\Delta ACD$ and $\Delta BCD$ are both isosceles triangles, we have $\overline{AF} \perp \overline{CD}$ and $\overline{BF} \perp \overline{CD}$. Thus, $\overline{CD}$ is perpendicular to the plane containing $\Delta ABF$.

By the Pythagorean Theorem, we have $AF = \frac{3}{5}\sqrt{17}$ and $BF = \frac{2}{5}\sqrt{82}$. By Heron's formula and some algebra, we have $[ABF] = 3\sqrt{2}$. Thus, the volume of tetrahedron $ABCD$ is the sum of the volumes of pyramids $CABF$ and $DABF$, which is $\frac{1}{3}[ABF] \cdot CD = (\text{C})\frac{24}{5}$.

## Exercises: 3-D Geometry

### Exercise 1.3.1

(AMC 10B 2018 Problem 10)

In the rectangular parallelepiped shown, $AB = 3$, $BC = 1$, and $CG = 2$. Point $M$ is the midpoint of $\overline{FG}$. What is the volume of the rectangular pyramid with base $BCHE$ and apex $M$?

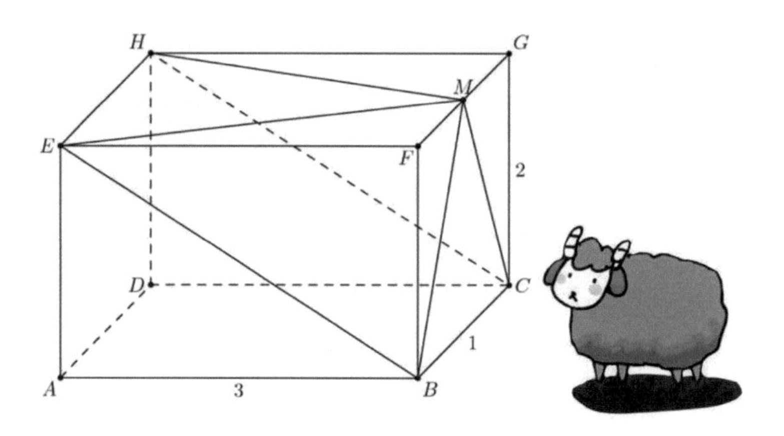


### Exercise 1.3.2

(AMC 10A 2019 Problem 21 & AMC 12A 2019 Problem 18)

A sphere with center O has radius 6. A triangle with sides of length 15, 15, and 24 is situated in space so that each of its sides is tangent to the sphere. What is the distance between O and the plane determined by the triangle?

## Exercise 1.3.3

(AMC 10A 2013 Problem 22 & AMC 12A 2013 Problem 18)

Six spheres of radius 1 are positioned so that their centers are at the vertices of a regular hexagon of side length 2. The six spheres are internally tangent to a larger sphere whose center is the center of the hexagon. An eighth sphere is externally tangent to the six smaller spheres and internally tangent to the larger sphere. What is the radius of this eighth sphere?

## Exercise 1.3.4

(AMC 10B 2012 Problem 23)

A solid tetrahedron is sliced off a solid wooden unit cube by a plane passing through two nonadjacent vertices on one face and one vertex on the opposite face not adjacent to either of the first two vertices. The tetrahedron is discarded and the remaining portion of the cube is placed on a table with the cut surface face down. What is the height of this object?

## Exercise 1.3.5

(AMC 10B 2014 Problem 23 & AMC 12B 2014 Problem 19)

A sphere is inscribed in a truncated right circular cone as shown. The volume of the truncated cone is twice that of the sphere. What is the ratio of the radius of the bottom base of the truncated cone to the radius of the top base of the truncated cone?

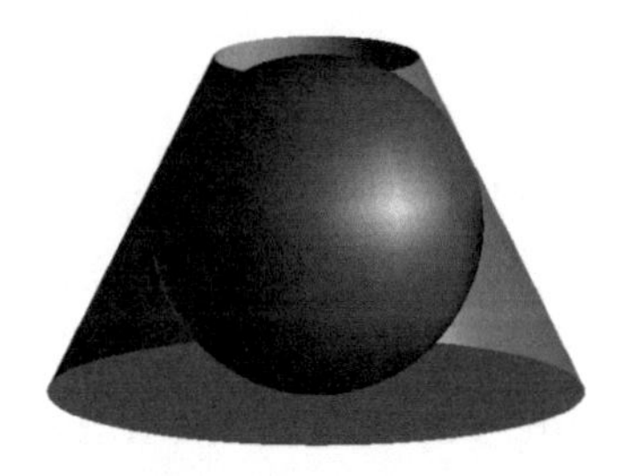

## Exercise 1.3.6

(AMC 12A 2011 Problem 15)

The circular base of a hemisphere of radius 2 rests on the base of a square pyramid of height 6. The hemisphere is tangent to the other four faces of the pyramid. What is the edge-length of the base of the pyramid?

## Exercise 1.3.7

(AMC 12B 2012 Problem 19)

A unit cube has vertices $P_1$, $P_2$, $P_3$, $P_4$, $P_1'$, $P_2'$, $P_3'$, and $P_4'$. Vertices $P_2$, $P_3$, and $P_4$ are adjacent to $P_1$, and for $1 \le i \le 4$, vertices $P_i$ and $P_i'$ are opposite to each other. A regular octahedron has one vertex in each of the segments $P_1P_2$, $P_1P_3$, $P_1P_4$, $P_1'P_2'$, $P_1'P_3'$, and $P_1'P_4'$. What is the octahedron's side length?

## Exercise 1.3.8

(AIME I 2017 Problem 4)

A pyramid has a triangular base with side lengths 20, 20, and 24. The three edges of the pyramid from the three corners of the base to the fourth vertex of the pyramid all have length 25. The volume of the pyramid is $m\sqrt{n}$, where m and n are positive integers, and n is not divisible by the square of any prime. Find m + n.

## Exercise 1.3.9

(Tiger)

A ball of radius 3 is dropped from a random point from the ceiling of a room shaped like a cube of side length 18. A right circular cone of radius $2\sqrt{3}$ and height 6 is placed at the center of the room. Find the probability that the ball hits cone before hitting the ground.

## Exercise 1.3.10
(AIME I 2015 Problem 15)

A block of wood has the shape of a right circular cylinder with radius 6 and height 8, and its entire surface has been painted blue. Points A and B are chosen on the edge of one of the circular faces of the cylinder so that $\widehat{AB}$ on that face measures $120°$. The block is then sliced in half along the plane that passes through point A, point B, and the center of the cylinder, revealing a flat, unpainted face on each half. The area of one of these unpainted faces is $a \cdot \pi + b\sqrt{c}$, where $a, b$, and $c$ are integers and $c$ is not divisible by the square of any prime. Find $a + b + c$.

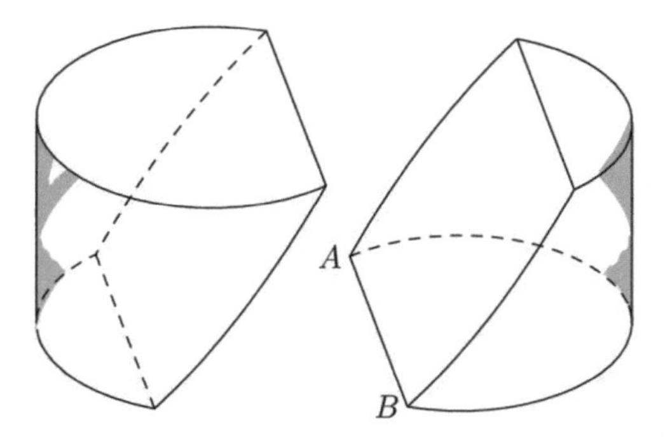


## Example 1.4
(AMC 10B 2019 Problem 23)

Points A(6, 13) and B(12, 11) lie on circle $\omega$ in the plane. Suppose that the tangent lines to $\omega$ at A and B intersect at a point on the $x$-axis. What is the area of $\omega$?

(A) $\frac{83\pi}{8}$ (B) $\frac{21\pi}{2}$ (C) $\frac{85\pi}{8}$ (D) $\frac{43\pi}{4}$ (E) $\frac{87\pi}{8}$

## Discussion

**KEVIN** 2019 was your first year trying the AMC 10. Do you remember if you solved this problem in contest?

**TIGER** I didn't get to it in time, but I think the best solution involves a technique I call "locus bounding." It consists of finding the locus of points that satisfy a particular condition and using the locus as the new condition.

We first notice that $TA = TB$, implying that $T$ must be on the perpendicular bisector of $\overline{AB}$. Now, instead of using the tangency condition or $TA = TB$, we only use the perpendicular bisector condition to find $T$. This is the mindset of locus bounding.

**KEVIN** I did this problem using a different approach. I found the coordinates of $T$ by solving a quadratic equation using $TA = TB$ and the distance formula.

**TIGER** That also works. To find the radius of the circle, we can use similar triangles to relate lengths we don't know to lengths we do know.

**KEVIN** For this part, I used the so-called Altitude-on-Hypotenuse Theorem, which is derived from the similar triangles.

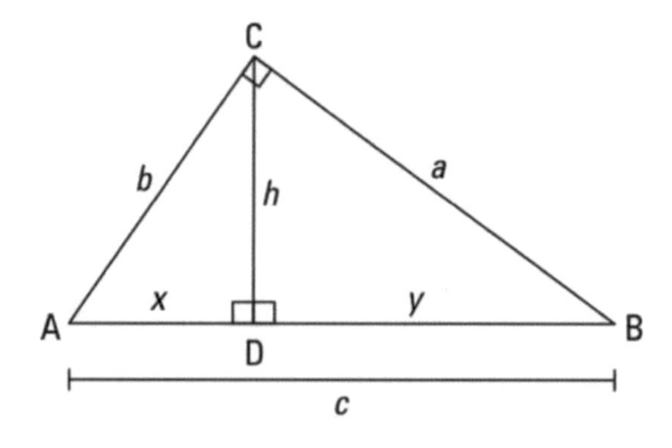

**ALTITUDE-ON-HYPOTENUSE THEOREM:** If an altitude is drawn to the hypotenuse of a right triangle as shown in the above figure, then

1. The two triangles formed are similar to the given triangle and to each other: $\Delta ABC \sim \Delta ADC \sim \Delta CDB$,
2. $h^2 = xy$, and
3. $a^2 = yc$ and $b^2 = xc$.

## Tiger's Solution

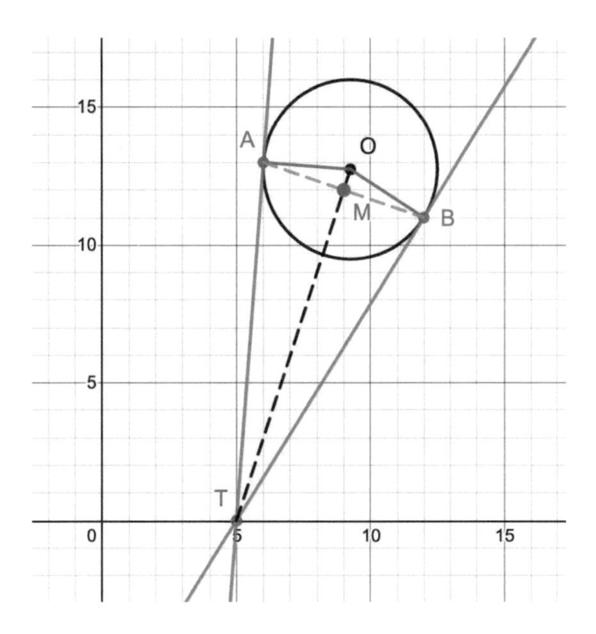


WWW.DESMOS.COM/CALCULATOR/XJJBPNUCMU

Let O be the center of circle $\omega$ and let $T$ be the intersection of the tangents to $\omega$ at $A$ and $B$. $\overline{TA}$ and $\overline{TB}$ are both tangents to $\omega$, so $TA = TB$. Thus, $T$ is on the perpendicular bisector of $\overline{AB}$.

Let $M$ be the midpoint of $\overline{AB}$, so $M = (9, 12)$. The slope of $\overline{AB}$ is $-\frac{1}{3}$, so the slope of $\overline{MT}$ is 3. Thus, the equation of line $MT$ is $y - 12 = 3(x - 9)$. It intersects the $x$-axis at $(5, 0)$, so $T = (5, 0)$.

By AA similarity, $\Delta AMT \sim \Delta OAT$. So $\frac{OA}{AT}=\frac{AM}{MT}$. It is easy to calculate $AT = \sqrt{170}$, $AM = \sqrt{10}$, and $MT = 4\sqrt{10}$, so $OA = \frac{\sqrt{170}}{4}$. Therefore, the area of $\omega$ is $\left(\frac{\sqrt{170}}{4}\right)^2 \pi = \text{(C)}\ \frac{85\pi}{8}$.

## Kevin's Solution

Let O be the center of the circle $\omega$ and let $T = (t, 0)$ be the intersection of the tangents to $\omega$ at $A$ and $B$. $\overline{AB}$ intersects $\overline{OT}$ at $M$. $\overline{TA}$ and $\overline{TB}$ are both tangents to $\omega$, so $TA = TB$.

By the distance formula, $(t-6)^2 + 13^2 = (t-12)^2 + 11^2$. This gives $t = 5$.

$\overline{TA}$ and $\overline{TB}$ are both tangents to $\omega$, so $TA = TB$. Thus, $T$ is on the perpendicular bisector of $\overline{AB}$. Let $M$ be the midpoint of $\overline{AB}$, so $M = (9, 12)$. By the Altitude-on-Hypotenuse Theorem, $AT^2 = TM \cdot TO$ and $OA^2 = OM \cdot OT$. It is easy to calculate $AT^2 = 170$, $TM = \sqrt{160}$. Therefore $TO = \frac{170}{\sqrt{160}}$, $OM = TO - TM = \frac{10}{\sqrt{160}}$, and $OA^2 = \frac{85}{8}$.

Thus, the area of $\omega$ is $\pi\, OA^2 = \text{(C)}\frac{85\pi}{8}$.

# Exercises: Locus Bounding

## Exercise 1.4.1

(Tiger)

Let $A$ and $B$ be distinct points. Find the set of all points $P$ such that $\angle APB = \theta$, where $0° < \theta < 180°$.

## Exercise 1.4.2
(Tiger)

Let $A = (-1, 0)$, $B = (-2,5)$, $C = (1, 0)$, and $D = (2, 2)$ on the coordinate plane. How many points $P$ exist such that $\Delta PAB$ and $\Delta PCD$ are both right triangles?

## Exercise 1.4.3
(MATHCOUNTS State 2021 Sprint Problem 29)

Marian throws a dart that lands randomly on a dartboard shaped like an isosceles trapezoid with side lengths 12 inches, 12 inches, 12 inches, and 24 inches. What is the probability that the dart is closer to the 24-inch side than it is to any of the other three sides of the dartboard?

## Exercise 1.4.4
(Tiger)

Points $A$ and $B$ and line $l$ lie on a plane such that the distance between $A$ and $l$ is $x$ and the distance between $B$ and $l$ is $y$. Given that there exists exactly one point $C$ on $l$ such that $\angle ACB = 90°$, find $AB$ in terms of $x$ and $y$.

## Exercise 1.4.5
(Tiger)

Let $a > 1$ be a real number, let $\omega_1$ be the circle defined by $x^2 + y^2 = a^2$, and let $\omega_2$ be the circle defined by $x^2 + (y - 2)^2 = a^2$. Suppose $A$ and $B$ are distinct points on $\omega_1$ such that their reflections across $(1, 1)$ lies on $\omega_2$. Find $AB$ in terms of $a$.

## Exercise 1.4.6

(AMC 10B 2015 Problem 19)

In $\Delta ABC$, $\angle C = 90°$ and $AB = 12$. Squares $ABXY$ and $ACWZ$ are constructed outside of the triangle. The points $X$, $Y$, $Z$, and $W$ lie on a circle. What is the perimeter of the triangle?

## Exercise 1.4.7

(Tiger)

$ABCD$ is a square with circumcenter O. Point $E$ is randomly and uniformly chosen inside $ABCD$. Let $\omega$ be the circle centered at O passing through $E$. Find the probability that exactly two of the segments $AE$, $BE$, $CE$, and $DE$ pass through the interior of $\omega$.

## Exercise 1.4.8

(HMMT February 2019 Geometry Problem 3)

Let $\overline{AB}$ be a line segment with length 2, and $S$ be the set of points $P$ on the plane such that there exists point $X$ on segment $\overline{AB}$ with $AX = 2PX$. Find the area of $S$.

## Exercise 1.4.9

(AIME I 2008 Problem 10)

Let $ABCD$ be an isosceles trapezoid with $\overline{AD} \parallel \overline{BC}$ whose angle at the longer base $\overline{AD}$ is $\frac{\pi}{3}$. The diagonals have length $10\sqrt{21}$, and point $E$ is at distances $10\sqrt{7}$ and $30\sqrt{7}$ from vertices $A$ and $D$, respectively. Let $F$ be the foot of the altitude from $C$ to $\overline{AD}$. The distance $EF$ can be expressed in the form $m\sqrt{n}$, where $m$ and $n$ are positive integers and $n$ is not divisible by the square of any prime. Find $m + n$.

## Exercise 1.4.10
(AIME II 2022 Problem 12)

Let $a$, $b$, $x$, and $y$ be real numbers with $a > 4$ and $b > 1$ such that

$$\frac{x^2}{a^2} + \frac{y^2}{a^2 - 16} = \frac{(x-20)^2}{b^2-1} + \frac{(y-11)^2}{b^2} = 1.$$

Find the least possible value of $a + b$.

## Example 1.5
(2013 AMC 10B Problem 23 & 2013 AMC 12B Problem 19)

In triangle $ABC$, $AB = 13$, $BC = 14$ and $CA = 15$. Distinct points $D$, $E$, and $F$ lie on segments $\overline{BC}$, $\overline{CA}$, and $\overline{DE}$, respectively, such that $\overline{AD} \perp \overline{BC}$, $\overline{DE} \perp \overline{AC}$, and $\overline{AF} \perp \overline{BF}$. The length of segment $\overline{DF}$ can be written as $\frac{m}{n}$, where $m$ and $n$ are relatively prime positive integers. What is $m + n$?

(A) 18 (B) 21 (C) 24 (D) 27 (E) 30

## Discussion

**KEVIN** When I was working on this problem, it took me an hour to realize that I had identified the wrong similar triangles. What was your approach?

**TIGER** We want to draw a good diagram. It is well known that a 13-14-15 triangle is split into a 5-12-13 and a 9-12-15 triangle by altitude $AD$. If we didn't know that, we could derive it by calculating the area of $\Delta ABC$ by Heron's Formula, then finding $AD$. Then, we could use the Pythagorean Theorem to find $BD$ and $CD$. We can get information about $E$ with similar triangles $\Delta AED \sim \Delta DEC$, but we hit a barrier when trying to find information about point $F$.

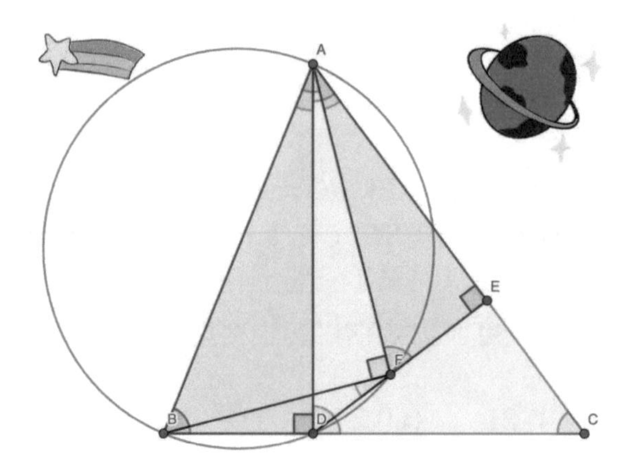


WWW.GEOGEBRA.ORG/CALCULATOR/BXYZ2GEV

We know that $\angle AFB = 90°$, so we can think about locus bounding to turn that condition into the condition that $F$ is on a circle with diameter $\overline{AB}$.

**KEVIN** Point $D$ is also on this circle, so $ABDF$ is cyclic. Cyclic quadrilaterals are powerful tools used in many AIME geometry problems and virtually all Olympiad geometry problems.

**TIGER** Yes, this produces a series of congruent angles:

$$\angle ADB = \angle AFB = \angle AED = 90^\circ$$
$$\angle ABF = \angle ADF = 90^\circ - \angle CDE = \angle ACD$$
$$\angle BAF = 180^\circ - \angle BDF = \angle CDF = 90^\circ - \angle ADE = \angle CAD$$
$$\angle ABD = 180^\circ - \angle AFD = \angle AFE$$
$$\angle BAD = \angle BFD = 90^\circ - \angle AFE = \angle EAF$$

Not all of these angle congruences will be used. However, we don't know which angle congruences will be useful, so it's best to mark in as many as we can.

**KEVIN** I often use the following less-than-well-known theorem to identify congruent angles: If two sides of an angle are each perpendicular to another angle's two sides, the two angles are congruent.

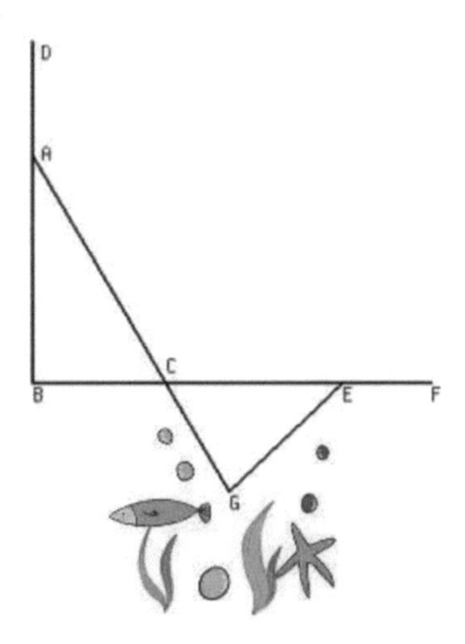


For example, in the above diagram

Given $\angle B = \angle G = 90^\circ$, then $\angle BAC$ has two sides $\overline{AB}$ and $\overline{AC}$; $\angle CEG$ has two sides $\overline{EC}$ and $\overline{EG}$; $\overline{AB} \perp \overline{EC}$ and $\overline{AC} \perp \overline{EG}$. Therefore $\angle BAC = \angle CEG$.

The proof is very simple:
Both angles are complementary to the vertical angles $\angle ACB = \angle ECG$. This theorem allows us to easily identify congruent angles that do not share a side and do not relate to any parallel lines.

**TIGER** Because we know $DE$, we can find $DF$ if we can find $EF$. We can look at the right triangle containing $\overline{EF}$, $\Delta AFE$. The only triangle that looks similar to $\Delta AFE$ is $\Delta ABD$. We use our previous angle congruences to prove these triangles are indeed similar by AA similarity.

This is a great angle chasing problem and demonstrates the power of the two most useful tools—similar triangles and cyclic quadrilaterals. I think of them as "strong" tools; they can take care of floating angles that are very hard to calculate using other methods. Similar triangles create length and area ratios, thus relating angles and lengths. Cyclic quadrilaterals come from the inscribed angle theorem, which is related to symmetry, a powerful tool in math. These tools are used more frequently in harder geometry problems, so familiarize yourself with them now.

## Tiger's Solution

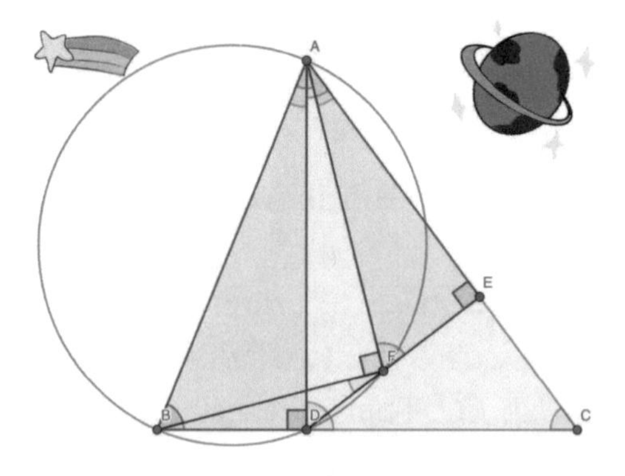


WWW.GEOGEBRA.ORG/CALCULATOR/BXYZ2GEV

We use Heron's Formula to find $[ABC] = \sqrt{21(21-13)(21-14)(21-15)}$ $= 84$.

Since $[ABC] = \frac{1}{2}BC \cdot AD$, we have $AD = 12$. By the Pythagorean Theorem, we have $BD = 5$ and $CD = 9$. Since $\Delta AED \sim \Delta DEC$ by AA, we can get $DE = \frac{36}{5}$ and $AE = \frac{48}{5}$.

Since $\angle ADB = \angle AFB = 90°$, $D$ and $F$ are on the circle with diameter $\overline{AB}$. Thus, we have $\angle BAD = \angle BFD$. We have $\angle BFD = 180° - 90° - \angle AFE =$ $90° - \angle AFE = \angle EAF$, so by AA similarity, $\Delta ABD \sim \Delta AFE$. Therefore, $\frac{FE}{AE} = \frac{BD}{AD} = \frac{5}{12}$, so $FE = \frac{5}{12} \cdot \frac{48}{5} = 4$ and $DF = DE - FE = \frac{16}{5}$, giving an answer of (**B**) 21.

## Exercises: Angle Chasing

### Exercise 1.5.1

(AMC 8 2016 Problem 23)

Two congruent circles centered at points $A$ and $B$ each pass through the other circle's center. The line containing both $A$ and $B$ is extended to intersect the circles at points $C$ and $D$. The circles intersect at two points, one of which is $E$. What is the degree measure of $\angle CED$?

### Exercise 1.5.2

(Tiger)

$ABCD$ is a parallelogram with $AB = 116$ and $BC = 73$. Suppose $E \neq D$ is on segment $\overline{CD}$ such that $AD = AE$. If $\angle BAC = \angle CAE$, find $DE$.

### Exercise 1.5.3
(Tiger)

Regular pentagons $ABEFG$ and $DCHIJ$ are constructed on a plane such that $ABCD$ is a square, $C$ is inside pentagon $ABEFG$, and $A$ is outside pentagon $DCHIJ$. What is $\angle AEI$?

### Exercise 1.5.4
(Tiger)

Let $ABCDEF$ be an equiangular hexagon with $AB = AF = DC = DF = 32$ and $BC = EF = 21$. The circumcircle of $\Delta ABD$ intersects $\overline{AF}$ at a point $P \neq A$. Find $FP$.

### Exercise 1.5.5
(AIME I 2020 Problem 1)

In $\Delta ABC$ with $AB = AC$, point $D$ lies strictly between $A$ and $C$ on side $\overline{AC}$, and point $E$ lies strictly between $A$ and $B$ on side $\overline{AB}$ such that $AE = ED = DB = BC$. The degree measure of $\angle ABC$ is $\frac{m}{n}$, where $m$ and $n$ are relatively prime positive integers. Find $m + n$.

### Exercise 1.5.6
(Tiger)

In $\Delta ABC$, let $D$ be a point on segment $\overline{BC}$. Let $E$ be the point on $\overline{AB}$ such that $\overline{AC} \parallel \overline{DE}$ and let $F$ be the point on $\overline{AC}$ such that $\overline{AB} \parallel \overline{DF}$. Let $M$ be the midpoint of $\overline{BC}$. Prove that $\angle BAD = \angle CAM$ if $BCFE$ is cyclic.

## Exercise 1.5.7

(AMC 10A 2014 Problem 22)

In rectangle $ABCD$, $AB = 20$ and $BC = 10$. Let $E$ be a point on $\overline{CD}$ such that $\angle CBE = 15^\circ$. What is $AE$?

## Exercise 1.5.8

(Tiger)

Points $A$ and $B$ are on the unit circle $\omega$ centered at O. Point $C \neq O$ is a point with $AC = BC = 1$. Let $T$ be a point on $\omega$ such that $\overline{CT}$ is tangent to $\omega$. Let $M$ be the midpoint of $AB$. If $TM \cdot AB = \frac{120}{169}$ and $T$ is on minor arc $\widehat{AB}$, what is $AB$?

## Exercise 1.5.9

(ARML 2021 Team Problem 8)

Circle $\Omega_1$ with radius 11 and circle $\Omega_2$ with radius 5 are externally tangent. Circle $\Gamma$ is internally tangent to both $\Omega_1$ and $\Omega_2$, and the centers of all three circles are collinear. Line $\iota$ is tangent to $\Omega_1$ and $\Omega_2$ at distinct points $D$ and $E$, respectively. Point $F$ lies on $\Gamma$ so that $FD < FE$ and $\angle DFE = 90^\circ$. Compute $\frac{DF}{DE}$.

## Exercise 1.5.10

(AIME II 2021 Problem 14)

Let $\Delta ABC$ be an acute triangle with circumcenter O and centroid $G$. Let $X$ be the intersection of the line tangent to the circumcircle of $\Delta ABC$ at $A$ and the line perpendicular to $GO$ at $G$. Let $Y$ be the intersection of lines $XG$ and $BC$. Given that the measures of $\angle ABC$, $\angle BCA$, and $\angle XOY$

are in the ratio 13:2:17, the degree measure of $\angle BAC$ can be written as $\frac{m}{n}$, where $m$ and $n$ are relatively prime positive integers. Find $m + n$.

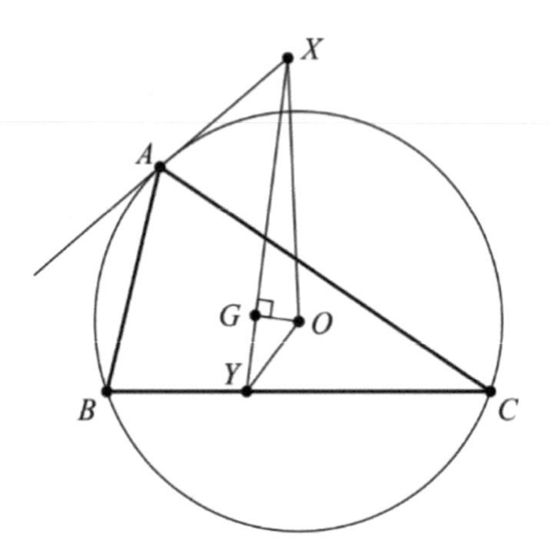


## Example 1.6

(AMC 10B Spring 2021 Problem 25 & AMC 12B Spring 2021 Problem 25)

Let $S$ be the set of lattice points in the coordinate plane, both of whose coordinates are integers between 1 and 30, inclusive. Exactly 300 points in $S$ lie on or below a line with equation $y = mx$. The possible values of $m$ lie in an interval of length $\frac{a}{b}$, where $a$ and $b$ are relatively prime positive integers. What is $a + b$?

(A) 31 (B) 47 (C) 62 (D) 72 (E) 85

## Discussion

**KEVIN** This was Problem 25 for both the Spring 2021 AMC 10B and 12B. Do you remember how much time you had left when you got to this problem?

**TIGER** I had about five minutes.

**KEVIN** And you solved this problem correctly in five minutes? When I tried this problem, it took me a while to realize that $m = \frac{2}{3}$ is the lower bound. How did you approach this problem?

**TIGER** There are 900 lattice points in $S$, and we want $y = mx$ to be above 300 of them. The problem is that lattice points are annoying to work with. So, I approximated the situation with areas. Consider a square with side length 30 and area 900 sitting on the $x$-axis and to the right of the $y$-axis. Then we want to cut the square into two pieces, one with area 300.

We know that $y = mx$ must pass through the origin. We let $y = mx$ pass through $(30, k)$. We know that $\frac{30 \cdot k}{2} = 300$, so $k = 20$. Thus, $m$ is roughly $\frac{2}{3}$.

**KEVIN** If we count the number of points below $y = mx$, then we get exactly 300.

**TIGER** If $m < \frac{2}{3}$, then some of the lattice points which were previously on $y = mx$ would be above it, so there would be less than 300 points. So, $m = \frac{2}{3}$ is the lower bound.

**KEVIN** It's harder to find the upper bound. I guessed that it should go through the point (30, 21) and got the wrong answer. (The wrong answer matched one of the answer choices, so it was anticipated by the problem writer.)

**TIGER** To find the upper bound, we think of increasing $m$ until the graph touches a lattice point, after which there are more than 300 points on or below $y = mx$. Because of the time constraints, I used an unrigorous argument.

**KEVIN** But it worked! Can you describe it?

**TIGER** I drew a few diagrams that were smaller than 30 by 30, like the one below. It looked like all the points that were above but closest to $y = \frac{2}{3}x$ had $x$-coordinates congruent to 1 mod 3 and were closest to the right boundary. A rigorous proof is written below.

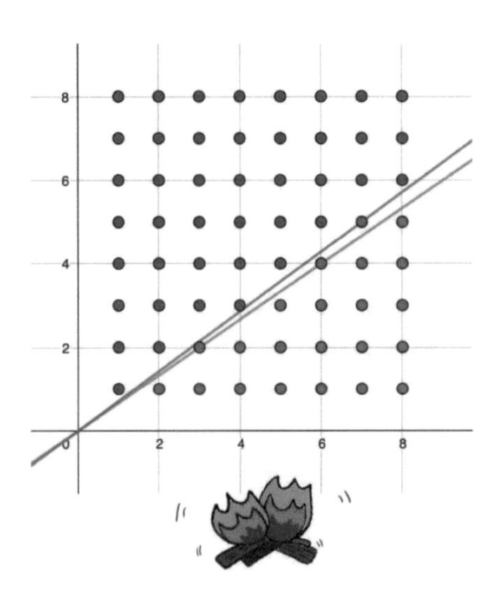


WWW.GEOGEBRA.ORG/CALCULATOR/RCQV2KHU

**KEVIN** The 30 by 30 diagram is like:

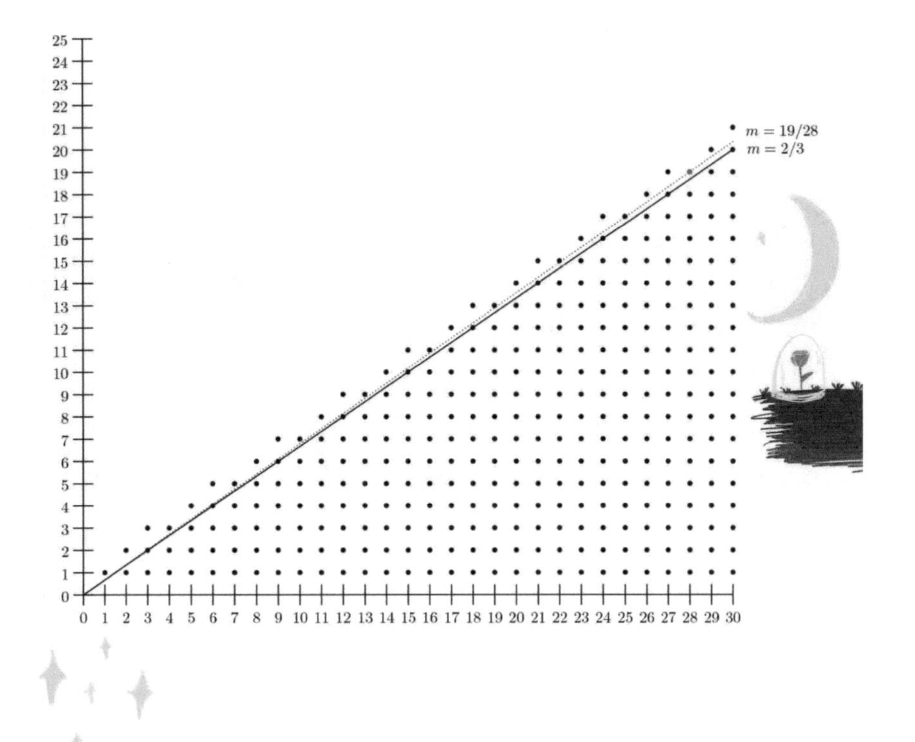


**(SOURCE: AOPS)**

**TIGER** For 30 by 30, $28 \equiv 1 \bmod 3$, so I picked $(28, 19)$ as the point that $y = m'x$ should go through with the upper bound $m' = \frac{19}{28}$.

The solution below is more rigorous.

## Tiger's Solution

First, we try $m = \frac{2}{3}$. There are $\left\lfloor \frac{2}{3}n \right\rfloor$ points on or below the graph of $y = mx$ at $x = n$. So, there are $(0+1+2)+(2+3+4)+(4+5+6)+\ldots+(18+19+20) = 3(1+3+5+\ldots+19) = 300$ points on or below the graph of $y = \frac{2}{3}x$. If $m < \frac{2}{3}$, then some of the points which were previously on $y = mx$ would be above it, so there would be less than 300 points. So, $m = \frac{2}{3}$ is the lower bound.

To find the higher bound, we are trying to find a fraction $\frac{p}{q}$ with positive integers $p, q \leq 30$ such that $\frac{p}{q} - \frac{2}{3}$ attains its least positive value. That expression is equivalent to $\frac{3p-2q}{3q}$. This is probably minimized when $3p - 2q = 1$, so we test that case. We find that $p \equiv 1 \bmod 2$ and $q \equiv 1 \bmod 3$ is the only case that works. Thus, $q = 28$ minimizes the expression if $3p - 2q = 1$, and that gives $p = 19$, which gives the answer of $\frac{1}{84}$. If $3p - 2q > 1$, then $\frac{3p-2q}{3q} \geq \frac{2}{3 \cdot 30} = \frac{1}{45}$, which is too large. Thus, the answer is (E) 85.

# Exercises: Lattice Points

### Exercise 1.6.1

(JMPSC Invitationals 2021 Problem 2, by Tiger)

Two quadrilaterals are drawn on the plane such that they share no sides. What is the maximum possible number of intersections of the boundaries of the two quadrilaterals?

## Exercise 1.6.2
(AMC 8 2004 Problem 14)

What is the area enclosed by the geoboard quadrilateral below?

## Exercise 1.6.3
(Tiger)

Let $A$, $B$, and $C$ be lattice points that form a nondegenerate obtuse triangle with $\angle ABC > 90°$. There exists a rectangle with one side of length $\frac{2}{15}$ that contains $A$, $B$, and $C$ in its boundary or interior. Find the least possible value of $AC$.

## Exercise 1.6.4
(AMC 8 2002 Problem 23)

A corner of a tiled floor is shown. If the entire floor is tiled in this way and each of the four corners looks like this one, then what fraction of the tiled floor is made of darker tiles?

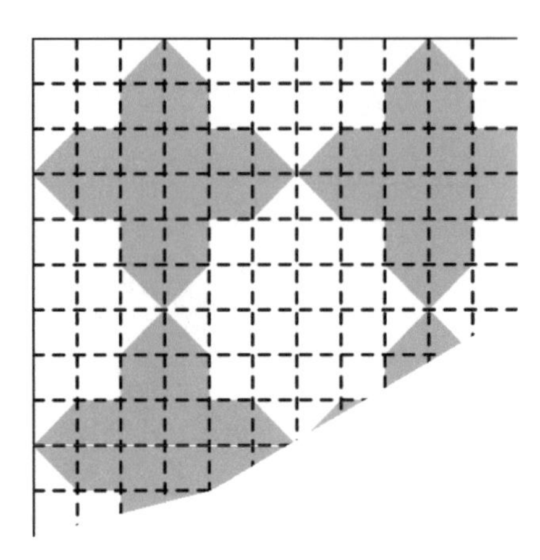

## Exercise 1.6.5
(AMC 10B 2011 Problem 24 & AMC 12B 2011 Problem 19)

A lattice point in an $xy$-coordinate system is any point $(x,y)$ where both $x$ and $y$ are integers. The graph of $y = mx + 2$ passes through no lattice point with $0 < x \leq 100$ for all $m$ such that $\frac{1}{2} < m < a$. What is the maximum possible value of $a$?

## Exercise 1.6.6
(AMC 12B 2009 Problem 22)

Parallelogram $ABCD$ has area 1,000,000. Vertex $A$ is at $(0,0)$ and all other vertices are in the first quadrant. Vertices $B$ and $D$ are lattice points on the lines $y = x$ and $y = kx$ for some integer $k > 1$, respectively. How many such parallelograms are there? (A lattice point is any point whose coordinates are both integers.)

## Exercise 1.6.7
(AMC 12B 2009 Problem 25)

The set $G$ is defined by the points $(x, y)$ with integer coordinates, $3 \le |x| \le 7,\ 3 \le |y| \le 7$. How many squares of side at least 6 have their four vertices in $G$?

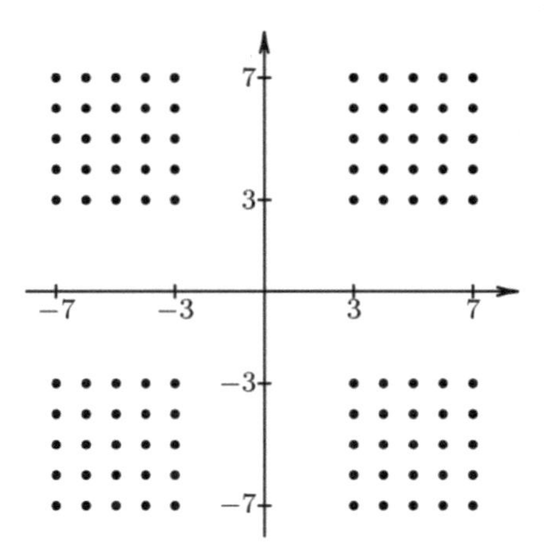


## Exercise 1.6.8
(AMC 10A 2022 Problem 25)

Let $R$, $S$, and $T$ be squares that have vertices at lattice points (i.e., points whose coordinates are both integers) in the coordinate plane, together with their interiors. The bottom edge of each square is on the $x$-axis. The left edge of R and the right edge of $S$ are on the $y$-axis, and $R$ contains $\frac{9}{4}$ as many lattice points as does $S$. The top two vertices of $T$ are in $R \cup S$, and $T$ contains $\frac{1}{4}$ of the lattice points contained in $R \cup S$. See the figure (not drawn to scale).

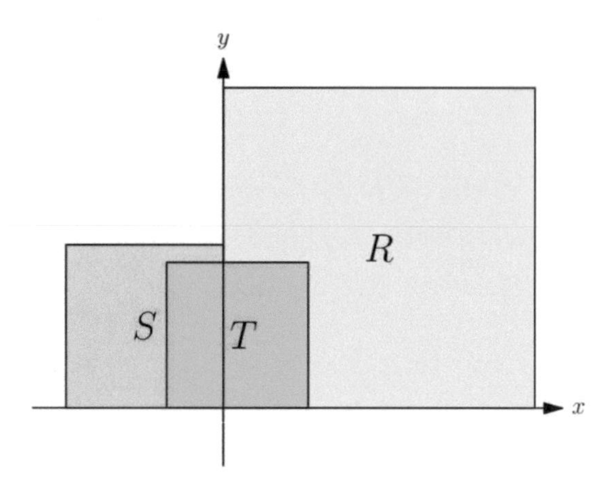


The fraction of lattice points in $S$ that are in $S \cap T$ is 27 times the fraction of lattice points in $R$ that are in $R \cap T$. What is the minimum possible value of the edge length of $R$ plus the edge length of $S$ plus the edge length of $T$?

## Exercise 1.6.9

(AIME I 2016 Problem 14)

Centered at each lattice point in the coordinate plane are a circle radius $\frac{1}{10}$ and a square with sides of length $\frac{1}{5}$ whose sides are parallel to the coordinate axes. The line segment from $(0, 0)$ to $(1001, 429)$ intersects $m$ of the squares and $n$ of the circles. Find $m + n$.

## Exercise 1.6.10

(IMO 1987 Problem 5)

Let $n$ be an integer greater than or equal to 3. Prove that there is a set of $n$ points in the plane such that the distance between any two points is irrational and each set of three points determines a non-degenerate triangle with rational area.

# *two*
# ALGEBRA

**KEVIN** Algebra is the foundation of mathematics. We all take Pre-algebra, Algebra 1, and Algebra 2 at school for many years. How do you feel when you see an algebra problem late in the AMC 10 contest?

**TIGER** Algebra is probably my worst subject, so I feel a little nervous when I see an algebra problem near the end.

**KEVIN** I feel the same. Algebra problems can be very hard, and sometimes I have no clue where to start. With geometry problems, on the other hand, it's usually easy to find a path to the solution.

**TIGER** The basics of algebra, like solving linear and quadratic equations, are important for many other math subjects.

**KEVIN** Algebra topics piqued my interest early in my math studies. When I was in fifth grade, my math teacher, Mr. Yu, taught us many new math subjects, including ways of solving system of equations. In the first two classes, he taught us how to solve linear equations by substitution and elimination. At the beginning of the third class, he told us, "This method may be the easiest. You can stare at the equations and find solutions by calculating mentally!" Then, Mr. Yu introduced the matrix method and Cramer's rule to fifth graders!

## Example 2.1
(AMC 10A 2015 Problem 23)

The zeroes of the function $f(x) = x^2 - ax + 2a$ are integers. What is the sum of all possible values of $a$?

(A) 7 (B) 8 (C) 16 (D) 17 (E) 18

### Discussion

**TIGER** There are a few common ways we can represent the fact that $r$ and $s$ are roots of a quadratic polynomial $P(x) = ax^2 + bx + c$:

1. $P(r) = P(s) = 0$; we plug in $r$ and $s$.
2. $r$ and $s$ follow the quadratic formula.
3. $P(x) = a(x - r)(x - s)$; we express $P(x)$ in terms of roots.
4. $r$ and $s$ satisfy Vieta's formulas.

**KEVIN** My first instinct is to use Vieta's formulas.

**VIETA'S FORMULA FOR QUADRATICS:**

Suppose a quadratic $ax^2 + bx + c$ has (not necessarily distinct) roots $r$ and $s$. Then,

$$r + s = -\frac{b}{a} \text{ and } rs = -\frac{c}{a}.$$

This theorem is used extensively in math competitions. Here is the version for degree $n$ polynomials:

**VIETA'S FORMULAS:**

Suppose a polynomial $a_n x^n + a_{n-1} x^{n-1} + \ldots + a_1 x + a_0$ has (not necessarily distinct) roots $r_1, r_2, \ldots, r_n$. For $k = 1, 2, \ldots, n$, define $s_k$ as the sum of the products of each combination of $k$ variables. Then, $s_k = \left(-1\right)^k \frac{a_{n-k}}{a_n}$.

By Vieta's formulas, we have:

$$r + s = a$$
$$rs = 2a$$

I concluded that $a$ must be an integer. But I couldn't continue with this strategy.

**TIGER** I went over the four strategies and concluded that plugging in variables doesn't seem to work. The quadratic formula works, but it turns out to be somewhat difficult. Between 3 and 4, 4 is more direct, so let's try that. (It actually doesn't matter whether we choose 3 or 4 because they're equivalent; 3 is used to derive 4.) Then, we can eliminate the variable $a$ and apply Simon's favorite factoring trick.

**KEVIN** Simon's favorite factoring trick[1] (SFFT): This trick solves Diophantine equations of the form $axy + bx + cy = d$. To solve this, multiply by $a$ on both sides to get $a^2xy + abx + acy = ad$, and notice that $(ax + c)(ay + b) = a^2xy + abx + acy + bc$. Then, we can add $bc$ to both sides to get $(ax + c)(ay + b) = ad + bc$, and this reduces to finding the factors of $ad + bc$ and matching them with values of $x$ and $y$.

**TIGER** We substitute the variable $a$ and obtain the equation $rs = 2r + 2s$. Using SFFT, this becomes $(r - 2)(s - 2) = 4$. Then, we find all pairs of integers that multiply to 4. These pairs are the possible values of $(r - 2, s - 2)$.

**KEVIN** I couldn't finish the Vieta's formulas path, so I tried the quadratic formula. The roots of the quadratic equation can be written in terms of $a$. For the roots to be integers, the discriminant must be a square number. Then, I factored it using difference of squares.

1 This method is named after Dr. Simon Rubinstein-Salzedo, a math researcher and educator who taught at AoPS between 2003 and 2005. However, it was known much earlier. (https://www.quora.com/How-did-you-discover-and-popularize-Simons-favorite-Factoring-Trick)

## Tiger's Solution

Let $r$ and $s$ be the two roots of $f(x)$. $r$ and $s$ are integers. By Vieta's formulas, $r + s = a$ and $rs = 2a$. We replace $r + s$ with $a$ to get $rs = 2r + 2s \Rightarrow rs - 2r - 2s = 0 \Rightarrow rs - 2r - 2s + 4 = (r - 2)(s - 2) = 4$. Thus, $r - 2$ and $s - 2$ are (possibly negative) factors of 4. Since the order of $r$ and $s$ doesn't matter, we can assume that $r \geq s$. We have the following possibilities:

$$r - 2 = 4, s - 2 = 1;$$
$$r - 2 = 2, s - 2 = 2;$$
$$r - 2 = -1, s - 2 = -4;$$
$$r - 2 = -2, s - 2 = -2.$$

These give the ordered pairs (6, 3), (4, 4), (1, –2), and (0, 0). We extract 9, 8, –1, and 0, respectively, as the possible values of $a$. Therefore, the answer is $9 + 8 - 1 + 0 =$ (C) 16.

## Kevin's Solution

By the quadratic formula, $f(x)$ has roots $x = \frac{a \pm \sqrt{a^2 - 8a}}{2}$. Since the roots are integers, $a$ must be an integer by Vieta's formulas. Then, we have $2x - a = \pm\sqrt{a^2 - 8a}$, so $\sqrt{a^2 - 8a}$ is an integer.

Let $\sqrt{a^2 - 8a} = n$ for a nonnegative integer $n$. We find $a^2 - 8a = n^2$, which rearranges to $(a - 4)^2 - n^2 = (a - 4 + n)(a - 4 - n) = 16$ after completing the square. Therefore, $a - 4 + n$ and $a - 4 - n$ must be factors of 16. Because $a - 4 + n \geq a - 4 - n$, and $a - 4 + n$ and $a - 4 - n$ are both odd or both even, we have these possibilities:

$$a - 4 + n = 8, a - 4 - n = 2;$$
$$a - 4 + n = 4, a - 4 - n = 4;$$
$$a - 4 + n = -2, a - 4 - n = -8;$$
$$a - 4 + n = -4, a - 4 - n = -4.$$

We add the two equations in each case and solve for $a$ to find that the only possible values of $a$ are 9, 8, –1, and 0, respectively. Therefore, the answer is $9 + 8 - 1 + 0 =$ (C) 16.

**TIGER** The main idea for my solution is to use Vieta's formulas to create an equation. The equation can be factored using Simon's favorite factoring trick, which should be kept in mind when finding integer solutions to an equation.

## Exercises: Polynomials and Simon's Favorite Factoring Trick

### Exercise 2.1.1

(AMC 10B 2002 Problem 6 & AMC 12B 2002 Problem 3)

For how many positive integers $n$ is $n^2 - 3n + 2$ a prime number?

### Exercise 2.1.2

(algebra.com)

Find all integers $x$ for which there exists an integer $y$ such that

$$\frac{1}{x}+\frac{1}{y}=\frac{1}{7}.$$

### Exercise 2.1.3
(HMMT February 2018 Algebra and Number Theory Problem 4)

Distinct prime numbers $p, q, r$ satisfy the equation

$$2pqr + 50pq = 7pqr + 55pr = 8pqr + 12qr = A$$

for some positive integer $A$. What is $A$?

### Exercise 2.1.4
(Tiger)

Let $a$, $b$, and $c$ be real numbers such that they are (not necessarily distinct) roots of $x^3 - ax^2 + bx + c = 0$. Prove that either $c = a$ or $c = b$. (This is harder than it seems! Make sure your work is organized and there are no division by 0 errors.)

### Exercise 2.1.5
(Tiger)

Find the number of degree 12 polynomials $P(x)$ such that $P(x)$ is positive if and only if $x$ is in $(-\infty, -9) \cup (-9, -2) \cup (2, 9) \cup (9, \infty)$. ( $(a, b)$ denotes the interval of real numbers between $a$ and $b$ not including $a$ or $b$, and $\cup$ denotes the union of intervals.)

### Exercise 2.1.6
(AMC 12B 2007 Problem 23)

How many non-congruent right triangles with positive integer leg lengths have areas that are numerically equal to 3 times their perimeters?

### Exercise 2.1.7
(AIME 1987 Problem 5)

Find $3x^2y^2$ if $x$ and $y$ are integers such that $y^2 + 3x^2y^2 = 30x^2 + 517$.

### Exercise 2.1.8
(AIME 1998 Problem 14)

An $m \times n \times p$ rectangular box has half the volume of an $(m + 2) \times (n + 2) \times (p + 2)$ rectangular box, where $m$, $n$, and $p$ are integers, and $m \leq n \leq p$. What is the largest possible value of $p$?

### Exercise 2.1.9
(BMO 2005 Round 2 Problem 1)

The integer $N$ is positive. There are exactly 2005 ordered pairs $(x, y)$ of positive integers satisfying:

$$\frac{1}{x}+\frac{1}{y}=\frac{1}{N}.$$

Prove that $N$ is a perfect square.

### Exercise 2.1.10
(IMO 1976 Problem 2)

Let $P_1(x) = x^2 - 2$ and let $P_j(x) = P_1\left(P_{j-1}(x)\right)$ for all integers $j \geq 2$. Prove that for any positive integer $n$, the roots of the equation $P_n(x) = x$ are all real and distinct.

## Example 2.2

(AMC 10A 2012 Problem 24 & AMC 12A 2012 Problem 21)

Let $a$, $b$, and $c$ be positive integers with $a \geq b \geq c$ such that

$$a^2 - b^2 - c^2 + ab = 2011$$

and

$$a^2 + 3b^2 + 3c^2 - 3ab - 2ac - 2bc = -1997$$

What is $a$?

(A) 249 (B) 250 (C) 251 (D) 252 (E) 253

## Discussion

**TIGER** This problem is quite difficult, but we've put it early in the chapter to demonstrate an important skill in algebra: algebraic manipulations.

We see a system of equations with 3 variables and 2 equations. In general, this doesn't uniquely determine the values of the variables, so will probably use the integer condition in some way. In a system of equations, it's almost always useful to add, subtract, or multiply two equations. If we add the two equations, we get $2a^2 + 2b^2 + 2c^2 - 2ab - 2bc - 2ca = 14$, or $a^2 + b^2 + c^2 - ab - bc - ca = 7$. This is good because the number on the right-hand side is small and the equation is symmetric.

Since the equation is of degree 2, we can write this as a sum of perfect squares. This has the advantage that squares are nonnegative. We can try $(a + b + c)^2$, but it only brings positive terms, not negative ones. If we want $-ab$, we can try $(a - b)^2$. That looks good, so we use symmetry and try $(a - b)^2 + (b - c)^2 + (a - c)^2 = 2a^2 + 2b^2 + 2c^2 - 2ab - 2bc - 2ca = 14$. Bingo! Now, we try to write 14 as a sum of perfect squares, and only $1 + 4 + 9$ works.

**KEVIN** It took me 30 minutes to solve this problem. I'm glad that I am a teacher now and no longer a math competitor! Looking at the five choices, I saw that they were all near 250. It would be hard to deal with such large solutions. But I guessed that the variables might be close to each other. So, I set up two new variables $s = a - b$ and $t = b - c$ to rewrite two equations and eliminate the third variable leaving an equation of $s$ and $t$ only. I ended up with an equation that had smaller coefficients. This new equation was much easier to solve.

**TIGER** Diophantine equations are hard to solve when the numbers involved are very big unless you find some way to factor. This motivates adding the two equations, which makes the right-hand side small. The motivation for writing $a^2 + b^2 + c^2 - ab - bc - ca$ can also come from the inequality $a^2 + b^2 + c^2 \geq ab + bc + ca$. Many inequalities can be written as a sum of squares, so we try that here.

## Tiger's Solution

We add the two equations to get

$$2a^2 + 2b^2 + 2c^2 - 2ab - 2bc - 2ca = 14,$$

which is equivalent to

$$(a-b)^2 + (b-c)^2 + (a-c)^2 = 14.$$

Since $a \geq b \geq c$, we have $a - c \geq a - b$, $a - c \geq b - c$. By inspection, the only possible ordered triples $(a-b, b-c, a-c)$ are $(1,2,3)$ and $(2,1,3)$.

If $a - b = 1$, the first equation becomes

$$a^2 - (a-1)^2 - (a-3)^2 + a(a-1) = 2011.$$

This can be simplified to $7a = 2021$, which has no integer solutions.

If $a - b = 2$, the first equation becomes

$$a^2 - (a-2)^2 - (a-3)^2 + a(a-2) = 2011.$$

This simplifies to $8a = 2024$, so the answer is $a =$ (E) 253.

## Kevin's Solution

Let $s = a - b$ and $t = b - c$. Note that $s$ and $t$ are both nonnegative integers.

Substitute $b = t + c$ and $a = s + t + c$ to both equations. After simplifying, we get

$$s^2 + t^2 + 3st + 3cs + 2ct = 2011,$$
$$s^2 + t^2 - st - 3cs - 2ct = -1977.$$

We add these two equations to get $s^2 + t^2 + st = 7$. By inspection, the only solutions are $(s, t) = (1, 2), (2, 1)$.

If $(s, t) = (1, 2)$, the first equation for $s$ and $t$ becomes $7c = 2000$, which has no integer solutions.

If $(s, t) = (2, 1)$, the first equation for $s$ and $t$ becomes $8c = 2000$, or $c = 250$, which gives an answer of $a = s + t + c =$ (E) 253.

## Exercises: Algebraic Manipulations

### Exercise 2.2.1

(AMC 10 2000 Problem 15 & AMC 12 2000 Problem 11)

Two non-zero real numbers, $a$ and $b$, satisfy $ab = a - b$. Which of the following is a possible value of $\frac{a}{b} + \frac{b}{a} - ab$?

(A) $-2$ (B) $\frac{-1}{2}$ (C) $\frac{1}{3}$ (D) $\frac{1}{2}$ (E) 2

### Exercise 2.2.2

(AMC 10A 2015 Problem 16)

If $y + 4 = (x - 2)^2$, $x + 4 = (y - 2)^2$, and $x \neq y$, what is the value of $x^2 + y^2$?

### Exercise 2.2.3
(Tiger)

Let a, b, c, and d be positive real numbers in an arithmetic progression in that order. If $a + b + c + d = 20$ and $abcd = 161$, what is $ab + cd$?

### Exercise 2.2.4
(Tiger)

Let $a_1, a_2, ..., a_{20}$ be a permutation of 1, 2, ..., 20 such that

$$2a_i \geq n + 1 - i$$

$1 \leq i \leq n$. What is the maximum value of $a_1 + 2a_2 + 3a_3 + ... + 20a_{20}$?

### Exercise 2.2.5
(AMC 10B 2003 Problem 24)

The first four terms in an arithmetic sequence are $x + y$, $x - y$, $xy$, and $\frac{x}{y}$, in that order. What is the fifth term?

### Exercise 2.2.6
(AMC 10B 2002 Problem 20)

Let $a$, $b$, and $c$ be real numbers such that $a - 7b + 8c = 4$ and $8a + 4b - c = 7$. Then what is $a^2 - b^2 + c^2$?

## Exercise 2.2.7

(AMC 12B 2013 Problem 17)

Let $a$, $b$, and $c$ be real numbers such that

$$a + b + c = 2, \text{ and } a^2 + b^2 + c^2 = 12.$$

What is the difference between the maximum and minimum possible values of $c$?

## Exercise 2.2.8

(AIME II 2014 Problem 5)

Real numbers $r$ and $s$ are roots of $p(x) = x^3 + ax + b$, and $r + 4$ and $s - 3$ are roots of $q(x) = x^3 + ax + b + 240$. Find the sum of all possible values of $|b|$.

## Exercise 2.2.9

(AIME II 2015 Problem 14)

Let $x$ and $y$ be real numbers satisfying $x^4y^5 + y^4x^5 = 810$ and $x^3y^6 + y^3x^6 = 945$. Evaluate $2x^3 + (xy)^3 + 2y^3$.

## Exercise 2.2.10

(HMMT February 2010 Algebra Problem 7)

Let $a$, $b$, $c$, $x$, $y$, and $z$ be complex numbers such that

$$a = \frac{b+c}{x-2}, \; b = \frac{c+a}{y-2}, \; c = \frac{a+b}{z-2}.$$

If $xy + yz + xz = 67$ and $x + y + z = 2010$, find the value of $xyz$.

## Example 2.3
(AMC 10A 2020 Problem 21)

There exists a unique strictly increasing sequence of nonnegative integers $a_1 < a_1 < \ldots a_k$ such that

$$\frac{2^{289}+1}{2^{17}+1} = 2^{a_1} + 2^{a_2} + \ldots + 2^{a_k}.$$

What is $k$?

(A) 117 (B) 136 (C) 137 (D) 273 (E) 306

## Discussion

**KEVIN** The left-hand side of this expression looks like the geometric series formula: for a general geometric sequence with leading term $a$ and common ratio $r \neq 1$, the sum of its first $n$ terms is

$$a + ar + \ldots + ar^{n-1} = \sum_{i=0}^{n-1} ar^i = a\left(\frac{1-r^n}{1-r}\right).$$

**TIGER** We can use the factorization $x^n - 1 = (x-1)(x^{n-1} + x^{n-2} + \cdots + x + 1)$ to prove the above geometric series formula.

If $|r| < 1$, we can "plug in" $n = \infty$ to get

$$a + ar + ar^2 + \ldots = \frac{a}{1-r}.$$

This is the infinite geometric series formula. A rigorous proof of this involves limits, a tool in calculus.

## Tiger's Solution

By the geometric series formula, we have

$$\frac{2^{289}+1}{2^{17}+1}=\frac{1+(2^{17})^{17}}{1+2^{17}}=(2^{17})^{16}-(2^{17})^{15}+\ldots-2^{17}+1$$
$$=(2^{16}+2^{15}+\ldots+1)((2^{17})^{15}+(2^{17})^{13}+\ldots+2^{17}+1).$$

We get rid of the negative terms by pairing them with a larger positive term:

$$((2^{17})^{16}-(2^{17})^{15})+((2^{17})^{14}-(2^{17})^{13})+\ldots+((2^{17})^{2}-2^{17})+1$$
$$=(2^{17}-1)((2^{17})^{15}+(2^{17})^{13}+\ldots+2^{17}+1).$$

Since $2^{17}-1=2^{16}+2^{15}+\ldots+1$, the expression becomes

$$(2^{16}+2^{15}+\ldots+1)((2^{17})^{15}+(2^{17})^{13}+\ldots+2^{17}+1).$$

Observe that when we expand this, no power of 2 appears more than once. Therefore, $k=17\cdot 8+1=$ (C)137.

# Exercises: Sequences and Series

### Exercise 2.3.1

(AMC 10B 2016 Problem 16)

The sum of an infinite geometric series is a positive number S, and the second term in the series is 1. What is the smallest possible value of S?

## Exercise 2.3.2

(AMC 10B 2013 Problem 21 & AMC 12B 2013 Problem 14)

Two non-decreasing sequences of nonnegative integers have different first terms. Each sequence has the property that each term beginning with the third is the sum of the previous two terms, and the seventh term of each sequence is N. What is the smallest possible value of N?

## Exercise 2.3.3

(AMC 10A 2012 Problem 22)

The sum of the first $m$ positive odd integers is 212 more than the sum of the first $n$ positive even integers. What is the sum of all possible values of $n$?

## Exercise 2.3.4

(Tiger)

Let $a$, $b$, and $c$ be nonnegative real numbers in an arithmetic progression, in that order. If they satisfy the system of equations

$$\sqrt{a} + \sqrt{b} + \sqrt{c} = 13,$$
$$\sqrt{ab} + \sqrt{bc} + \sqrt{ca} = 47.$$

What is $abc$?

## Exercise 2.3.5

(AMC 10A 2019 Problem 23)

Travis has to babysit the terrible Thompson triplets. Knowing that they love big numbers, Travis devises a counting game for them. First Tadd will say the number 1, then Todd must say the next two numbers (2 and 3), then Tucker must say the next three numbers (4, 5, 6), then Tadd must say the next four numbers (7, 8, 9, 10), and the process continues to rotate through the three children in order, each saying one more number than the previous child did, until the number 10,000 is reached. What is the 2019th number said by Tadd?

## Exercise 2.3.6

(AMC 10A 2014 Problem 24)

A sequence of natural numbers is constructed by listing the first 4, then skipping one, listing the next 5, skipping 2, listing 6, skipping 3, and, on the $n$th iteration, listing $n + 3$ and skipping $n$. The sequence begins 1, 2, 3, 4, 6, 7, 8, 9, 10, 13. What is the 500,000th number in the sequence?

## Exercise 2.3.7

(HMMT February 2017 Algebra Problem 2)

Find the value of

$$\sum_{1 \le a < b < c} \frac{1}{2^a 3^b 5^c}$$

(i.e. the sum of $\frac{1}{2^a 3^b 5^c}$ over all triples of positive integers $(a, b, c)$ satisfying $a < b < c$).

## Exercise 2.3.8
(AMC 10A 2018 Problem 25)

For a positive integer $n$ and nonzero digits $a$, $b$, and $c$, let $A_n$ be the $n$-digit integer each of whose digits is equal to $a$, let $B_n$ be the $n$-digit integer each of whose digits is equal to $b$; and let $C_n$ be the $2n$-digit (not $n$-digit) integer each of whose digits is equal to $c$. What is the greatest possible value of $a + b + c$ for which there are at least two values of $n$ such that $C_n - B_n = A_n^2$?

## Exercise 2.3.9
(OMO Fall 2015-2016 Problem 9)

Let $s_1, s_2, \ldots$ be an arithmetic progression of positive integers. Suppose that $s_{s_1} = x + 2$, $s_{s_2} = x^2 + 18$, and $s_{s_3} = 2x^2 + 18$. Determine the value of $x$.

## Exercise 2.3.10
(AIME I 2020 Problem 8)

A bug walks all day and sleeps all night. On the first day, it starts at point O faces east, and walks a distance of 5 units due east. Each night the bug rotates $60°$ counterclockwise. Each day it walks in this new direction half as far as it walked the previous day. The bug gets arbitrarily close to the point P. Then $OP^2 = \frac{m}{n}$, where $m$ and $n$ are relatively prime positive integers. Find $m + n$.

## Example 2.4
(AMC 10B 2018 Problem 25)

Let $\lfloor x \rfloor$ denote the greatest integer less than or equal to $x$. How many real numbers $x$ satisfy the equation $x^2 + 10,000\lfloor x \rfloor = 10,000x$?

(A) 197 (B) 198 (C) 199 (D) 200 (E) 201

## Discussion

**KEVIN** How do you approach problems involving the floor function $\lfloor x \rfloor$?

**TIGER** I often see what happens if you ignore the floor functions. This often gives you an approximate idea of the problem. Unfortunately, this doesn't work well in this problem. However, we see that 10000 is a large number, so we would want to put both of the 10000s on the right-hand side to get $x^2 = 10000x - 10000\lfloor x \rfloor$ , or $x^2 = 10000(x - \lfloor x \rfloor)$. Now, notice that $x - \lfloor x \rfloor$ is the fractional part of $x$, or $\{x\}$, so we have $x^2 = 10000\{x\}$.

It's easy to find the solutions using graphing. I analyzed the equation $x^2 = 25\{x\}$ by graphing and found 9 solutions. It's easy to generalize this and conclude that $x^2 = 10000\{x\}$ has 199 solutions.

**KEVIN** I used a setup that works well with floor function problems: Let $n = \lfloor x \rfloor$ and $a = \{x\}$, and notice that $x = n + a$. The given equation becomes an equation of $n$ and $a$. Then, we use the fact that $n$ is an integer and $0 \le a < 1$ to solve the resulting equation.

## Tiger's Solution

The given equation is equivalent to $x^2 = 10000x - 10000\lfloor x \rfloor$. Since $x - \lfloor x \rfloor = \{x\}$, we have $x^2 = 10000\{x\}$.

Since $0 \le \{x\} < 1$, we have $0 \le x^2 < 10000$, so $-100 < x < 100$, then $-100 \le \lfloor x \rfloor \le 99$. Then we will find the intersection points between graphs $y = x^2$ and $y = 10000\{x\}$. The following shows the graphs of $y = x^2$ and $y = 25\{x\}$.

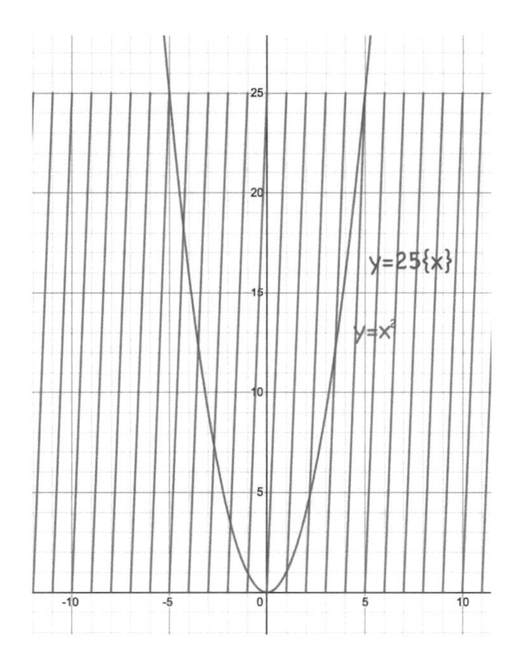


WWW.DESMOS.COM/CALCULATOR/EEKOFGN6XG

Notice that for any possible values of $\lfloor x \rfloor$ between -100 and 98, there is one value of $x$ such that $x^2 = 10000\{x\}$. For $\lfloor x \rfloor = 99$, we solve $x^2 = 10000(x - 99)$ to get $x = 100$ or $9900$, which contradicts $\lfloor x \rfloor = 99$. Thus, there are a total of (**C**) 199 solutions.

## Kevin's Solution

Let $x = n + a$, where $n$ is an integer and $0 \le a < 1$. So $n = \lfloor x \rfloor$. The equation becomes

$$(n + a)^2 + 10000n = 10000(n + a).$$

This simplifies to

$$a^2 - (10000 - 2n)a + n^2 = 0.$$

We use the quadratic formula to solve for $a$:

$$a = \frac{10000 - 2n \pm \sqrt{(10000-2n)^2 - 4n^2}}{2} = \frac{10000 - 2n \pm \sqrt{10000^2 - 40000n}}{2}$$
$$= 5000 - n \pm 100\sqrt{2500 - n}.$$

Thus, $n \le 2500$. Since $a < 1$, we must have – in place of ±, so

$$5000 - n - 100\sqrt{2500 - n} < 1 \Longrightarrow 100\sqrt{2500 - n} > 4999 - n.$$

We square both sides to get

$$n^2 + 2n - 9999 < 0 \Longrightarrow (n + 1)^2 < 10000 \Longrightarrow -100 \le n \le 98.$$

Thus, there are a total of **(C)** 199 solutions.

# Exercises: Floor Functions

## Exercise 2.4.1

Calculate

$$\lfloor \sqrt{1} \rfloor + \lfloor \sqrt{2} \rfloor + \lfloor \sqrt{3} \rfloor + \ldots + \lfloor \sqrt{2021} \rfloor.$$

## Exercise 2.4.2
(Mathematics Education)

Prove

$$\lfloor e \rfloor^{\lfloor \pi \rfloor} + \lfloor \pi \rfloor = \lfloor \pi \rfloor^{\lfloor e \rfloor} + \lfloor e \rfloor.$$

## Exercise 2.4.3
(Tiger)

Let $f(x) = \left\lfloor \frac{x}{1} \right\rfloor + \left\lfloor \frac{x}{2} \right\rfloor + \ldots$. Find the least positive integer $n$ such that $f(n+1) - f(n) = 10$.

## Exercise 2.4.4
(AMC 10B 2020 Problem 24 & AMC 12B 2020 Problem 21)

How many positive integers $n$ satisfy

$$\frac{n + 1000}{70} = \lfloor \sqrt{n} \rfloor ?$$

(Recall that $\lfloor x \rfloor$ is the greatest integer not exceeding $x$.)

## Exercise 2.4.5
(AMC 10B 2016 Problem 25)

Let

$$f(x) = \sum_{k=2}^{10} \left( \lfloor kx \rfloor - k \lfloor x \rfloor \right)$$

where $\lfloor r \rfloor$ denotes the greatest integer less than or equal to $r$. How many distinct values does $f(x)$ assume for $x \geq 0$?

## Exercise 2.4.6
(AMC 10A 2020 Problem 22)

For how many positive integers $n \leq 1000$ is $\left\lfloor \frac{998}{n} \right\rfloor + \left\lfloor \frac{999}{n} \right\rfloor + \left\lfloor \frac{1000}{n} \right\rfloor$ not divisible by 3? (Recall that $\lfloor x \rfloor$ is the greatest integer less than or equal to $x$.)

## Exercise 2.4.7
(AMC 10B 2015 Problem 21)

Cozy the Cat and Dash the Dog are going up a staircase with a certain number of steps. However, instead of walking up the steps one at a time, both Cozy and Dash jump. Cozy goes two steps up with each jump (though if necessary, he will just jump the last step). Dash goes five steps up with each jump (though if necessary, he will just jump the last steps if there are fewer than 5 steps left). Suppose that Dash takes 19 fewer jumps than Cozy to reach the top of the staircase. Let S denote the sum of all possible numbers of steps this staircase can have. What is the sum of the digits of S?

## Exercise 2.4.8
(AIME 1985 Problem 10)

How many of the first 1000 positive integers can be expressed in the form

$$\lfloor 2x \rfloor + \lfloor 4x \rfloor + \lfloor 6x \rfloor + \lfloor 8x \rfloor,$$

where $x$ is a real number, and $\lfloor z \rfloor$ denotes the greatest integer less than or equal to $z$?

## Exercise 2.4.9

(Tiger)

Let f be a function that maps nonnegative integers to nonnegative integers. If $f(0) = 0$ and

$$f(x) = f(x - \lfloor \sqrt{x} \rfloor) + x$$

for all nonnegative integers $x$, find $f(500)$.

## Exercise 2.4.10

(AIME I 2015 Problem 14)

For each integer $n \geq 2$, let $A(n)$ be the area of the region in the coordinate plane defined by the inequalities $1 \leq x \leq n$ and $0 \leq y \leq x\lfloor\sqrt{x}\rfloor$, where $\lfloor\sqrt{x}\rfloor$ is the greatest integer not exceeding $\sqrt{x}$. Find the number of values of $n$ with $2 \leq n \leq 1000$ for which $A(n)$ is an integer.

# *three*
# COUNTING AND PROBABILITY

**KEVIN** We start learning math by counting positive integers. As numbers become large and conditions become tricky, counting could become extremely complicated. We also encounter probability in our daily lives. Many probability problems involve counting techniques. Both counting and probability are part of a field in mathematics known as combinatorial analysis.

Counting and probability problems are not usually covered until Algebra 2 in high school or even later, and the scope covered is limited. However, such problems are common at all levels of math competitions. Tiger, how do you feel when you come across a late counting and probability problem in the AMC 10?

**TIGER** I think the AMC's counting and probability problems are of two types, casework problems and tricky problems. The casework problems are time-consuming but don't require much insight. The tricky problems can be quicker, but only if you see what to do.

**KEVIN** You're right. These problems may take a long time to solve, and the AMC has strict time limits. If such a problem appears in the AMC 10 problems 21-25—often at least two problems in this range are counting and probability problems—there is very little time to solve them.

I think this is the type of problem that is most likely to be answered incorrectly. In particular, an AIME counting problem has plenty of room for error because the answer must be an integer. Unless the answer is over 999 (all AIME problems have integer solutions between 0 and 999, inclusive), it's hard to know whether you have counted correctly or not. The AMC has multiple choice problems. You should feel "lucky" if you get an answer that is not one of the five choices, since you know you got it wrong. Then you can figure out if you have missed any cases.

**TIGER** I've developed the following list of strategies for counting and probability.

1. Knowledge of permutations and combinations and their applications
2. Casework
3. Constructive counting
4. Complementary counting
5. Principle of Inclusion and Exclusion (PIE)

6. One-to-one correspondences/bijections
7. Recursion
8. Basic probability principles
9. Analysis of states
10. Geometric probability
11. Expected value and linearity of expectation
12. Recognition of common problems (such as stars and bars and Catalan numbers situations)

Expected value is mostly an AIME topic, so we don't have to worry about it right now.

## Example 3.1

(AMC 10A 2021 Problem 25)

How many ways are there to place 3 indistinguishable red chips, 3 indistinguishable blue chips, and 3 indistinguishable green chips in the squares of a $3 \times 3$ grid so that no two chips of the same color are directly adjacent to each other, either vertically or horizontally?

(A) 12 (B) 18 (C) 24 (D) 30 (E) 36

## Discussion

**KEVIN** You did the AMC 10A 2021 a few months ago. Do you remember if you did this problem?

**TIGER** I got it wrong because I missed a few cases.

**KEVIN** How would you have solved the problem if you included all cases?

## Tiger's Solution

Without loss of generality (WLOG), we assume the center square contains a red chip.

We already know that the four squares adjacent to the center square don't contain red chips. We need to place the two remaining red chips in two of the four corner squares. There are two ways we can do this up to rotation and/or reflection: The two chips will either be on the same side or the same main diagonal.

**CASE 1:** The corner red chips are on the same side. We can assume WLOG that the bottom center square contains a blue chip. Then, the only possibility would be the figure on the right.

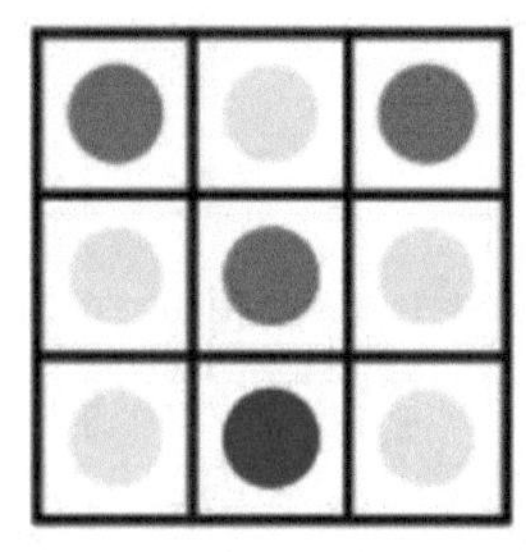

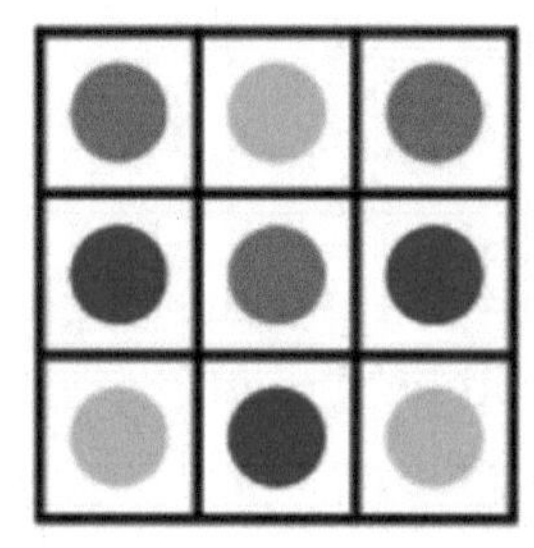

DIAGRAM SOURCE: ARTOFPROBLEMSOLVING.COM/COMMUNITY/C5H2440235P20216412

There are four possibilities if we account for rotation and reflection.

**CASE 2:** The corner red chips are on the same main diagonal. We can assume WLOG that the bottom left square contains a blue chip. If so, the figure on the right is the only possibility.

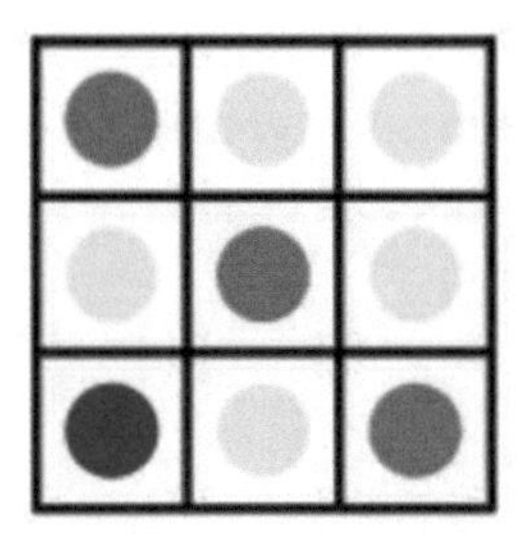
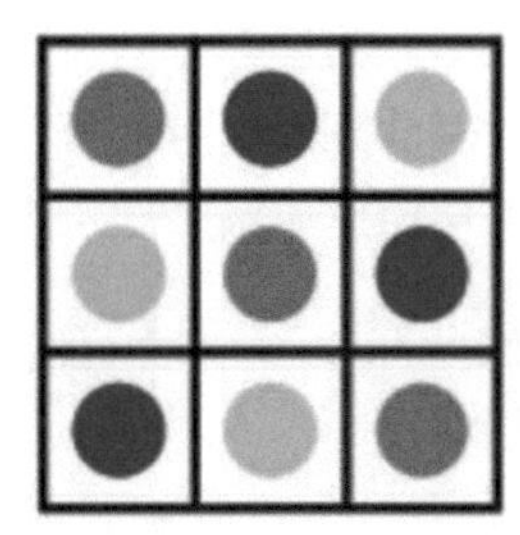

There are two possibilities if we account for rotation and reflection.

The WLOG assumptions we made about the first two chip colors reduced the number of possibilities by a factor of $3 \cdot 2 = 6$ (three possibilities for the first color and two possibilities for the second color). Thus, there are $6(4 + 2) =$ (D) 36 total possibilities.

## Kevin's Solution

Without loss of generality, we assume the first two squares in the top row contain a red chip and a blue chip, respectively. Then, we will multiply our count $3 \cdot 2 = 6$ to get the final answer.

We study the first two squares on the second row. There are three ways we can place the chips: blue and red, blue and green, and green and red.

**CASE 1:** The first two squares of the second row are blue and red. The remaining five squares must have three green chips. They must be separated so they have to be placed in the three corners. Then we have no place to put the third red chip.

| R | B | |
|---|---|---|
| B | R | |
| | | |

There is no solution in this case.

**CASE 2:** The first two squares of the second row are blue and green. The third blue chip can be put into squares (2, 3), (3, 2), or (3, 3).

| | | |
|---|---|---|
| R | B | |
| B | G | |
| | | |

There are three possibilities in this case:

| | | |
|---|---|---|
| R | B | R |
| B | G | B |
| G | R | G |

| | | |
|---|---|---|
| R | B | G |
| B | G | R |
| R | B | G |

| | | |
|---|---|---|
| R | B | G |
| B | G | R |
| G | R | B |

**CASE 3:** The first two squares of the second row are green and red. The third red chip can be put in squares (1, 3), (3, 1), or (3, 3).

| | | |
|---|---|---|
| R | B | |
| G | R | |
| | | |

There are three possibilities in this case:

| | | |
|---|---|---|
| R | B | R |
| G | R | G |
| B | G | B |

| | | |
|---|---|---|
| R | B | G |
| G | R | B |
| R | B | G |

| | | |
|---|---|---|
| R | B | G |
| G | R | B |
| B | G | R |

Again, the WLOG assumptions that we made on the first two chip colors reduced the number of possibilities by a factor of $3 \cdot 2 = 6$ (three possibilities for the first reduction and two possibilities for the second). Hence, there are $6(3 + 3) =$ (E)36 total possibilities.

**KEVIN** Your approach is very interesting. Using a WLOG argument, you set the middle square as red. I set the top left one to be red. Although both of them work, your approach is cleaner.

**TIGER** I tried using your approach during the competition. A common strategy for counting problems is to make WLOG assumptions, then base your constructive counting on these assumptions. In this problem, we demonstrate two important ideas: assuming a square contains a particular chip color and WLOGing up to rotations and reflections. Also, we use our WLOG argument on the center square, the most connected square, to get free information about it.

## Exercises: Constructive Counting and Casework

### Exercise 3.1.1

(AMC 8 2018 Problem 19)

In a sign pyramid a cell gets a "+" if the two cells below it have the same sign, and it gets a "−" if the two cells below it have different signs. The diagram below illustrates a sign pyramid with four levels. How many possible ways are there to fill the four cells in the bottom row to prduce a "+" at the top of the pyramid?

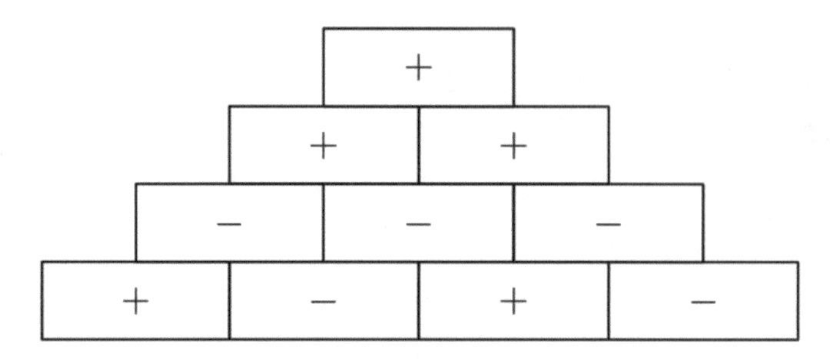


## Exercise 3.1.2

(AMC 10A 2022 Problem 14)

How many ways are there to split the integers 1 through 14 into 7 pairs such that in each pair, the greater number is at least 2 times the lesser number?

## Exercise 3.1.3

(AMC 10A 2021 Problem 15)

Values for $A$, $B$, $C$, and $D$ are to be selected from $\{1, 2, 3, 4, 5, 6\}$ without replacement (i.e. no two letters have the same value). How many ways are there to make such choices so that the two curves $y = Ax^2 + B$ and $y = Cx^2 + D$ intersect? (The order in which the curves are listed does not matter; for example, the choices $A = 3, B = 2, C = 4, D = 1$ is considered the same as the choices $A = 4, B = 1, C = 3, D = 2$.)

## Exercise 3.1.4
(AMC 10A 2012 Problem 23 & AMC 12A 2012 Problem 19)

Adam, Benin, Chiang, Deshawn, Esther, and Fiona have internet accounts. Some, but not all, of them are internet friends with each other, and none of them has an internet friend outside this group. Each of them has the same number of internet friends. In how many different ways can this happen?

## Exercise 3.1.5
(Tiger)

Let $N$ be the number of subsets of $\{1, 2, ..., 20\}$ such that no two elements add up to 20. What is the remainder when $N$ is divided by 5?

## Exercise 3.1.6
(AIME I 2016 Problem 3)

A regular icosahedron is a 20-faced solid where each face is an equilateral triangle and five triangles meet at every vertex. The regular icosahedron shown below has one vertex at the top, one vertex at the bottom, an upper pentagon of five vertices all adjacent to the top vertex and all in the same horizontal plane, and a lower pentagon of five vertices all adjacent to the bottom vertex and all in another horizontal plane. Find the number of paths from the top vertex to the bottom vertex such that each part of a path goes downward or horizontally along an edge of the icosahedron, and no vertex is repeated.

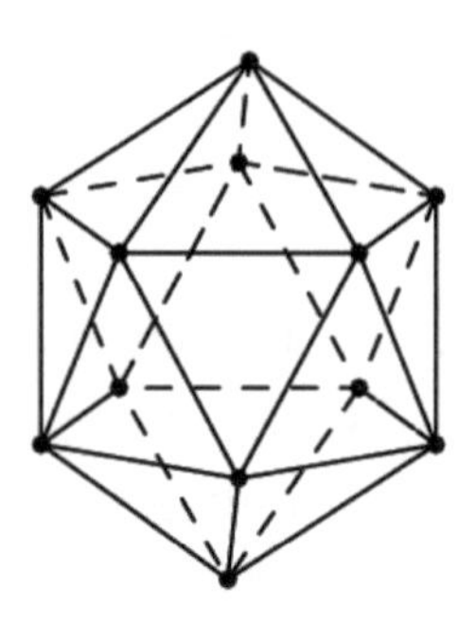

### Exercise 3.1.7

(AMC 10B 2012 Problem 24 & AMC 12B 2012 Problem 16)

Amy, Beth, and Jo listen to four different songs and discuss which ones they like. No song is liked by all three. Furthermore, for each of the three pairs of the girls, there is at least one song liked by those two girls but disliked by the third. In how many different ways is this possible?

### Exercise 3.1.8

(AMC 10A 2022 Problem 24 & AMC 12A 2022 Problem 24)

How many strings of length 5 formed from the digits 0, 1, 2, 3, 4 are there such that for each $j \in \{1, 2, 3, 4\}$, at least $j$ of the digits are less than $j$? (For example, 02214 satisfies this condition because it contains at least 1 digit less than 1, at least 2 digits less than 2, at least 3 digits less than 3, and at least 4 digits less than 4. The string 23404 does not satisfy the condition because it does not contain at least 2 digits less than 2.)

### Exercise 3.1.9

(AIME II 2010 Problem 8)

Let $N$ be the number of ordered pairs of nonempty sets $\mathscr{A}$ and $\mathscr{B}$ that have the following properties:

- $\mathscr{A} \cup \mathscr{B} = \{1, 2, 3, 4, 5, 6, 7, 8, 9, 10, 11, 12\}$,
- $\mathscr{A} \cap \mathscr{B} = \varnothing$,
- The number of elements of $\mathscr{A}$ is not an element of $\mathscr{A}$,
- The number of elements of $\mathscr{B}$ is not an element of $\mathscr{B}$.

Find $N$.

## Exercise 3.1.10

(Tiger)

A team of 10 people are planning to take a math competition. The competition consists of two tests for algebra, geometry, combinatorics, number theory, and calculus, and each teammate is supposed to do one test. Each person is good at two subjects and no two people are good at the same two subjects. Find the number of ways the teammates can choose the tests they will take if each person plans to take a test in a subject they're great at.

## Example 3.2

(AMC 10B 2016 Problem 22)

A set of teams held a round-robin tournament in which every team played every other team exactly once. Every team won 10 games and lost 10 games; there were no ties. How many sets of three teams {A, B, C} were there in which A beat B, B beat C, and C beat A?

(A) 385 (B) 665 (C) 945 (D) 1140 (E) 1330

## Discussion

**TIGER** My solution to this problem uses complementary counting, the idea of counting the total then subtracting the unwanted cases.

**KEVIN** The Principle of Inclusion and Exclusion (PIE) is a viable strategy in some complementary counting problems. First, we have to define some notation: let $A \cup B$ denote the set of elements in either $A$ or $B$, let $A \cap B$ denote the set of elements in both $A$ and $B$, and let $|A|$ denote the number of elements in $A$. For two sets $A$ and $B$, PIE states that

$$|A \cup B| = |A| + |B| - |A \cap B|.$$

For three sets $A$, $B$, and $C$, PIE states that

$$|A \cup B \cup C| = |A| + |B| + |C| - |A \cap B| - |B \cap C| - |A \cap C| + |A \cap B \cap C|.$$

**TIGER** The motivation behind doing complementary counting in this problem is that the condition we are given is very difficult to work with. However, its complementary statement turns out to be quite nice. In most of the easier complementary counting problems, it's clear that complementary counting would help when we see the words "not" or "at least" in the problem statement. However, the complementary counting solution in this problem is encoded. Even if you think of complementary counting, it isn't immediately clear how to proceed. This makes the problem very interesting and tricky.

## Tiger's Solution

Since each team won 10 games and lost 10 games, each team played 20 games. Therefore, there are 21 teams in total.

We can use a diagram to illustrate the situation: we denote each team as a point and draw an arrow for each game from the winning team to the losing team. The condition that A beats B, B beats C, and C beats A can be represented by a cycle from A to B to C and back to A.

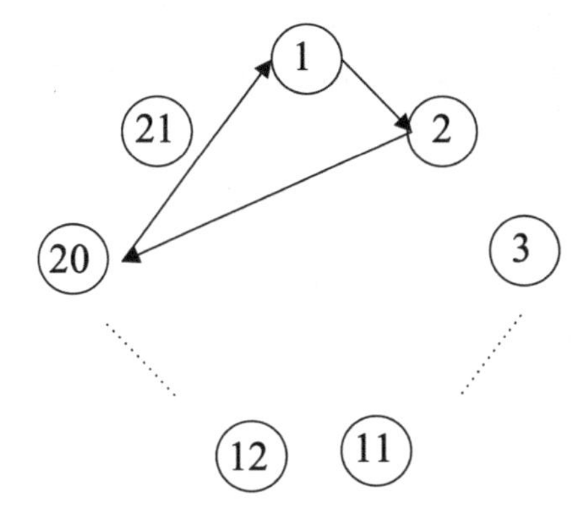


We count the number of three-team sets that don't form a cycle. In such a set, there must be exactly one team that beats the other two teams. Since each team won 10 games, there are $\binom{10}{2} = 45$ ways to choose two teams it beats. For 21 teams, there are $21 \cdot 45 = 945$ non-cycle sets. The total number of sets of three teams is $\binom{21}{3} = 1330$. Thus, there are $1330 - 945 =$ (**A**) 385 cycles.

**TIGER** In my solution, I found that counting non-cycles is much easier than counting cycles, so I used complementary counting. If direct counting seems difficult, you can always try complementary counting.

Complementary counting is a type of overcounting, in which we count more than we are supposed to, then subtract the extra cases. In some cases, it's also useful to overcount then divide by a constant. For example, suppose we want to find the number of ways to put 5 distinct beads on a necklace, where rotations and reflections of a necklace are equivalent. We can see that there are $5! = 120$ ways to put the beads on the necklace without considering rotations and reflections. However, there are 10 ways to rotate/reflect the necklace, so we overcounted by a factor of 10. Therefore, the answer is $120/10 = 12$.

## Exercises: Complementary Counting and PIE

### Exercise 3.2.1

(Tiger)

How many permutations of $(1, 2, 3, 4, 5)$ are there such that no three adjacent numbers add up to 10?

### Exercise 3.2.2

(AMC 8 2017 Problem 24)

Mrs. Sanders has three grandchildren, who call her regularly. One calls her every three days, one calls her every four days, and one calls her every five days. All three called her on December 31, 2016. On how many days during the next year did she not receive a phone call from any of her grandchildren?

## Exercise 3.2.3
(Tiger)

A school has 850 students and teachers combined. Each teacher teaches exactly 120 students and each student is taught by exactly 5 teachers. How many students are there in the school?

## Exercise 3.2.4
(AMC 10B 2022 Problem 12)

A pair of fair 6-sided dice is rolled $n$ times. What is the least value of $n$ such that the probability that the sum of the numbers face up on a roll equals 7 at least once is greater than $\frac{1}{2}$?

## Exercise 3.2.5
(AMC 10B 2022 Problem 18)

Consider systems of three linear equations with unknowns $x$, $y$, and $z$,

$$\begin{aligned} a_1x + b_1y + c_1z &= 0 \\ a_2x + b_2y + c_2z &= 0 \\ a_3x + b_3y + c_3z &= 0 \end{aligned}$$

where each of the coefficients is either 0 or 1 and the system has a solution other than $x = y = z = 0$. For example, one such system is

$$\{1x + 1y + 0z = 0, 0x + 1y + 1z = 0, 0x + 0y + 0z = 0\}$$

with a nonzero solution of $\{x, y, z\} = \{1, -1, 1\}$. How many such systems of equations are there? (The equations in a system need not be distinct, and two systems containing the same equations in a different order are considered different.)

### Exercise 3.2.6

(AMC 10A 2017 Problem 23)

How many triangles with positive area have all their vertices at points $(i, j)$ in the coordinate plane, where $i$ and $j$ are integers between 1 and 5, inclusive?

### Exercise 3.2.7

(AMC 12A 2010 Problem 18)

A 16-step path is to go from $(-4, -4)$ to $(4, 4)$ with each step increasing either the $x$-coordinate or the $y$-coordinate by 1. How many such paths stay outside or on the boundary of the square $-2 \leq x \leq 2$, $-2 \leq y \leq 2$ at each step?

### Exercise 3.2.8

(AIME I 2022 Problem 6)

Find the number of ordered pairs of integers $(a, b)$ such that the sequence

$$3, 4, 5, a, b, 30, 40, 50$$

is strictly increasing and no set of four (not necessarily consecutive) terms forms an arithmetic progression.

### Exercise 3.2.9
(AIME I 2020 Problem 5)

Six cards numbered 1 through 6 are to be lined up in a row. Find the number of arrangements of these six cards where one of the cards can be removed leaving the remaining five cards in either ascending or descending order.

### Exercise 3.2.10
(AIME II 2009 Problem 6)

Let $m$ be the number of five-element subsets that can be chosen from the set of the first 14 natural numbers so that at least two of the five numbers are consecutive. Find the remainder when $m$ is divided by 1000.

## Example 3.3
(AMC 10A 2011 Problem 21)

Two counterfeit coins of equal weight are mixed with 8 identical genuine coins. The weight of each of the counterfeit coins is different from the weight of each of the genuine coins. A pair of coins is selected at random without replacement from the 10 coins. A second pair is selected at random without replacement from the remaining 8 coins. The combined weight of the first pair is equal to the combined weight of the second pair. What is the probability that all 4 selected coins are genuine?

(A) $\frac{7}{11}$ (B) $\frac{9}{13}$ (C) $\frac{11}{15}$ (D) $\frac{15}{19}$ (E) $\frac{15}{16}$

## Discussion

**TIGER** Problems that have conditions like this are called conditional probability problems. Many of them can be solved by carefully using basic rules of probability.

**KEVIN** A good summary of the conditional probability can be found at https://www.investopedia.com/terms/c/conditional_probability.asp.

**TIGER** When studying cases, we have to be careful about making sure that each outcome occurs with equal probability. For example, suppose there are three coins: one in which both sides are black, one in which both sides are red, and one in which there is one black side and one red side. If one of the coins is flipped and the red side comes up, what is the probability that the other side is also red?

**KEVIN** The condition forces the coin to be one of the coins with a red side. But under this condition, these two coins don't appear with equal probability.

**TIGER** If we naively assume that both red and black occur with the same probability on the other side, then we would get an answer of $1/2$. However, the right answer would be $2/3$ because two of the three red sides belong to the coin with two red sides.

## Tiger's Solution

The probability is the ratio between the number of good outcomes and the number of total outcomes. The number of good outcomes is the number of possibilities there are of selecting 4 genuine coins when choosing two pairs, which is $\binom{8}{2}\binom{6}{2} = 420$.

To find the number of total outcomes, we have two cases: either all coins are genuine or there is one counterfeit coin in both pairs. There are $2 \cdot 8 \cdot 1 \cdot 7 = 112$ ways for both pairs to have one counterfeit coin. Thus, the number of total outcomes is $420 + 112 = 532$.

It follows that the probability is $\frac{420}{532} = \frac{15}{19}$, giving the answer of (**D**).

## Kevin's Solution

Let $A$ be the number of counterfeit coins selected in the first pair and let $B$ be the number of counterfeit coins selected in the second pair. The conditional probability the problem asks for is then $P(A = 0 \,\&\, B = 0 \mid A = B)$.

By simple probability calculations, we have

$$P(A = 0) = \frac{\binom{8}{2}}{\binom{10}{2}} = \frac{28}{45}, \qquad P(A = 1) = \frac{\binom{2}{1}\binom{8}{1}}{\binom{10}{2}} = \frac{16}{45},$$

$$P(A = 2) = \frac{\binom{2}{2}}{\binom{10}{2}} = \frac{1}{45}, \qquad P(B = 0 \mid A = 0) = \frac{\binom{6}{2}}{\binom{8}{2}} = \frac{15}{28},$$

$$P(B = 1 \mid A = 1) = \frac{\binom{7}{1}}{\binom{8}{2}} = \frac{1}{4}, \qquad P(B = 2 \mid A = 2) = 0.$$

By the conditional probability formula $P(X \,\&\, Y) = P(X)P(Y|X)$, we have

$$P(A=B=0) = P(A=0 \,\&\, B=0) = P(A=0)\,P(B=0\,|A=0\big) = \frac{28}{45}\cdot\frac{15}{28} = \frac{1}{3},$$

$$P(A=B=1) = P(A=1 \,\&\, B=1) = P(A=1)\,P(B=1\,|A=1) = \frac{16}{45}\cdot\frac{1}{4} = \frac{4}{45},$$

$$P(A=B=2) = P(A=2 \,\&\, B=2) = P(A=2)\,P(B=2\,|A=2) = \frac{1}{45}\cdot 0 = 0.$$

Therefore,

$$P(A=0 \,\&\, B=0 \mid A=B) = \frac{P(A=0 \,\&\, B= \,\&\, A=B)}{P(A=B)}$$

$$= \frac{P(A=B=0)}{P(A=B=0)+P(A=B=1)+P(A=B=2)} = \frac{\frac{1}{3}}{\frac{1}{3}+\frac{4}{45}+0} = \frac{15}{19}.$$

Therefore, the answer is (**D**).

## Exercises: Basic and Conditional Probability

### Exercise 3.3.1

(AMC 12B 2010 Problem 18)

A frog makes 3 jumps, each exactly 1 meter long. The directions of the jumps are chosen independently at random. What is the probability that the frog's final position is no more than 1 meter from its starting position?

## Exercise 3.3.2

(Tiger)

Carrie has a deck of 13 cards numbered from 1 to 13. She removes three of the cards randomly. Then, she shuffles the remaining cards thoroughly and takes the top two cards, which are 1 and 2, in no particular order. She puts these two cards back in the deck. After she thoroughly shuffles cards again, what is the probability that the top two cards are 3 and 4, in no particular order?

## Exercise 3.3.3

(MATHCOUNTS State Target Round 2019 Problem 7)

Andy has a cube of edge length 10 cm. He paints the outside of the cube red and then divides the cube into smaller cubes, each of edge length 1 cm. Andy randomly chooses one of the unit cubes and rolls it on a table. If the cube lands so that an unpainted face is on the bottom, touching the table, what is the probability that the entire cube is unpainted?

## Exercise 3.3.4

(Generalized Monty Hall Problem)

In a game show, a contestant chooses one room out of $n$ closed rooms, $k$ of which contain cars and the rest of which are empty. After the contestant chooses a closed room, the host chooses a random empty room and opens it. Then, the host gives the contestant a chance to change their mind and choose another room. If they do so, what is the probability that they choose a room with a car? (Assume $0 \leq k \leq n - 2$.)

## Exercise 3.3.5

(AMC 10B 2021 Fall Problem 20)

In a particular game, each of 4 players rolls a standard 6-sided die. The winner is the player who rolls the highest number. If there is a tie for the highest roll, those involved in the tie will roll again and this process will continue until one player wins. Hugo is one of the players in this game. What is the probability that Hugo's first roll was a 5 given that he won the game?

## Exercise 3.3.6

(AMC 12B 2020 Problem 20)

Two different cubes of the same size are to be painted, with the color of each face being chosen independently and at random to be either black or white. What is the probability that after they are painted, the cubes can be rotated to be identical in appearance?

## Exercise 3.3.7

(AMC 12A 2016 Problem 23)

Three numbers in the interval $[0, 1]$ are chosen independently and at random. What is the probability that the chosen numbers are the side lengths of a triangle with positive area?

## Exercise 3.3.8

(AIME II 2014 Problem 2)

Arnold is studying the prevalence of three health risk factors, denoted by A, B, and C, within a population of men. For each of the three factors, the probability that a randomly selected man in the population has only this risk factor (and none of the others) is $0.1$. For any two of the three factors, the probability that a randomly selected man has exactly these two risk factors (but not the third) is $0.14$. The probability that a randomly selected man has all three risk factors, given that he has A and B is $\frac{1}{3}$. The probability that a man has none of the three risk factors given that he does not have risk factor A is $\frac{p}{q}$, where $p$ and $q$ are relatively prime positive integers. Find $p + q$.

### Exercise 3.3.9
(AIME 1989 Problem 5)

When a certain biased coin is flipped five times, the probability of getting heads exactly once is not equal to 0 and is the same as that of getting heads exactly twice. Let $\frac{i}{j}$, in lowest terms, be the probability that the coin comes up heads in exactly 3 out of 5 flips. Find $i + j$.

### Exercise 3.3.10
(AIME I 2011 Problem 12)

Six men and some number of women stand in a line in random order. Let $p$ be the probability that a group of at least four men stand together in the line, given that every man stands next to at least one other man. Find the least number of women in the line such that $p$ does not exceed 1 percent.

## Example 3.4
(AMC 10B 2019 Problem 22 & AMC 12B 2019 Problem 19)

Raashan, Sylvia, and Ted play the following game. Each starts with \$1. A bell rings every 15 seconds, at which time each of the players who currently have money simultaneously chooses one of the other two players independently and at random and gives $1 to that player. What is the probability that after the bell has rung 2019 times, each player will have $1? (For example, Raashan and Ted may each decide to give $1 to Sylvia, and Sylvia may decide to give her dollar to Ted, at which point Raashan will have $0, Sylvia will have $2, and Ted will have $1, and that is the end of the first round of play. In the second round Rashaan has no money to give, but Sylvia and Ted might choose each other to give their $1 to, and the holdings will be the same at the end of the second round.)

(A) $\frac{1}{7}$ (B) $\frac{1}{4}$ (C) $\frac{1}{3}$ (D) $\frac{1}{2}$ (E) $\frac{2}{3}$

## Discussion

**TIGER** It's natural to ask about the possible distributions of money.

**KEVIN** We use the word "state" to describe the condition of a system. In this problem, a state is a possible distribution of money among three people. For example, we can identify ($2, $1, $0) as a state in which one person has $2, one has $1, and the other has no money. For simplicity, we don't need to identify which person has $2 and which person has $1. In a state diagram, we draw an arrow between two states and label it with the probability of going from one state to the other.

## Tiger's Solution

The total amount of money is \$3, so the money distribution must be (\$1, \$1, \$1), (\$2, \$1, \$0), or (\$3, \$0, \$0). However, it's impossible to get (\$3, \$0, \$0) since each person with money must give away \$1 at every round, so nobody can have all the money at the end of a round. We have two cases.

**CASE 1:** the distribution is (\$1, \$1, \$1). Then, the only way the distribution can stay at (\$1, \$1, \$1) is if everyone gives \$1 and receives \$1, and there are two ways of doing that. Thus, the probability of staying at (\$1, \$1, \$1) is $\frac{1}{4}$, and the probability of going to (\$2, \$1, \$0) is $\frac{3}{4}$.

**CASE 2:** the distribution is (\$2, \$1, \$0). Then, the only way the distribution can change to (\$1, \$1, \$1) is if the person with \$2 gives \$1 to the person with \$1, and the person with \$1 gives \$1 to the person with \$0. Thus, the probability of going to (\$1, \$1, \$1) is $\frac{1}{4}$ and the probability of staying at (\$2, \$1, \$0) is $\frac{3}{4}$.

Here's the state diagram, with calculation of probabilities of transiting from one state to itself or to another state:

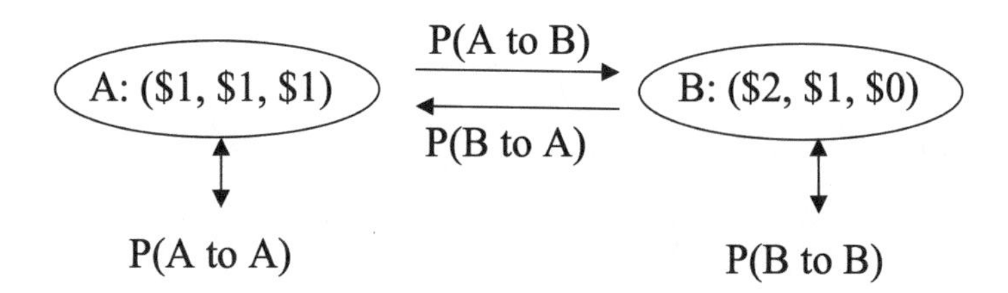


After any round, the probability of having a (\$1, \$1, \$1) distribution is always $\frac{1}{4}$, so the final probability is **(B)** $\frac{1}{4}$.

## Kevin's Solution

From the state diagram:

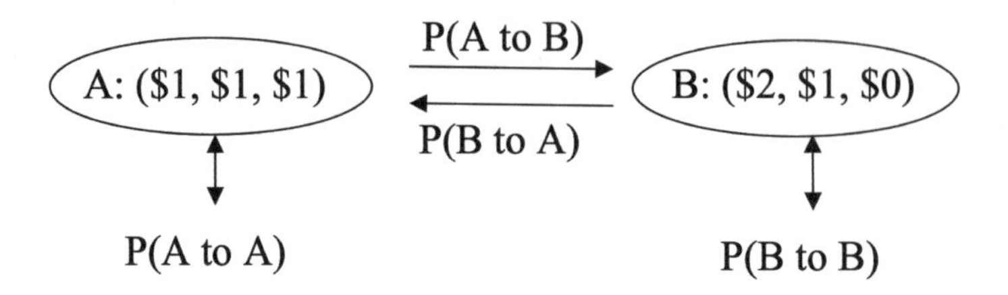


Let $p = P(A \text{ to } A)$, $q = P(B \text{ to } B)$, so $P(A \text{ to } B) = 1 - p$, $P(B \text{ to } A) = 1 - q$.

Let $P_n$ be the probability of arriving at state $A$ after $n$ rounds, and $Q_n$ be the probability of arriving at state $B$ after $n$ rounds. The problem asks for $P_{2019}$.

We know
$P_0 = 1, Q_0 = 0;\ P_1 = P(A \text{ to } A) = p,\ Q_1 = P(A \text{ to } B) = 1 - p.$

In general, for $n \geq 0$,
$P_{n+1} = P_n P(A \text{ to } A) + Q_n P(B \text{ to } A) = pP_n + (1-q)\,Q_n,$
$Q_{n+1} = P_n P(A \text{ to } B) + Q_n P(B \text{ to } B) = (1-p)\,P_n + qQ_n.$

In this problem, we have $p = \frac{1}{4}$ and $q = \frac{3}{4}$. So $P_0 = 1$, $Q_0 = 0$; $P_1 = \frac{1}{4}$, $Q_1 = \frac{3}{4}$.
$P_{n+1} = \frac{1}{4}P_n + \frac{1}{4}Q_n$, $Q_{n+1} = \frac{3}{4}P_n + \frac{3}{4}Q_n$.

Therefore, $3P_{n+1} = Q_{n+1}$ for $n \geq 0$, thus $P_{n+1} = \frac{1}{4}P_n + \frac{3}{4}P_n = P_n$ for $n \geq 1$.
So $P_n = P_1 = \frac{1}{4}$ for $n \geq 1$.

Therefore, the answer is **(B)** $\frac{1}{4}$.

**TIGER** For this problem, once the probability of each transition is calculated, the final answer becomes obvious. We got lucky when we found that the probability of going to (\$1, \$1, \$1) is $\frac{1}{4}$ regardless of the situation. Sometimes, the situation won't be that simple. In a case where we have different probabilities in different situations, we can make a sequence $a_n$, which represents the probability of being in a state, and find a recursive formula for $a_n$.

**KEVIN** Typically solving recursive equations is quite time-consuming, so they rarely appear on the AMC. However, they do appear on the AIME and other computational contests.

## Exercises: States

### Exercise 3.4.1

(AMC 10B 2018 Problem 6)

A box contains 5 chips, numbered 1, 2, 3, 4, and 5. Chips are drawn randomly one at a time without replacement until the sum of the values drawn exceeds 4. What is the probability that 3 draws are required?

### Exercise 3.4.2

(AMC 10A 2020 Problem 13 & AMC 12A 2020 Problem 11)

A frog sitting at the point (1, 2) begins a sequence of jumps, where each jump is parallel to one of the coordinate axes and has length 1, and the direction of each jump (up, down, right, or left) is chosen independently at random. The sequence ends when the frog reaches a side of the square with vertices (0, 0), (0, 4), (4, 4), and (4, 0). What is the probability that the sequence of jumps ends on a vertical side of the square?

## Exercise 3.4.3

(HMMT February 2017 Combinatorics Problem 4)

Sam spends his days walking around the following $2 \times 2$ grid of squares.

| | |
|---|---|
| 1 | 2 |
| 4 | 3 |

Say that two squares are adjacent if they share a side. He starts at the square labeled 1 and every second walks to an adjacent square. How many paths can Sam take so that the sum of the numbers on every square he visits in his path is equal to 20 (not counting the square he started on)?

## Exercise 3.4.4

(AMC 10B 2019 Problem 21)

Debra flips a fair coin repeatedly, keeping track of how many heads and how many tails she has seen in total, until she gets either two heads in a row or two tails in a row, at which point she stops flipping. What is the probability that she gets two heads in a row but she sees a second tail before she sees a second head?

## Exercise 3.4.5

(Tiger)

An ant is walking on a grid with two rows and three columns. It starts at the top left square. Every minute, it walks to an adjacent square (a square that shares a side with the square it was on before). Find the probability that the ant will reach the bottom left square before it reaches the bottom right square.

## Exercise 3.4.6

(AMC 10B 2014 Problem 25 & AMC 12B 2014 Problem 22)

In a small pond there are eleven lily pads in a row labeled 0 through 10. A frog is sitting on pad 1. When the frog is on pad $N$, $0 < N < 10$, it will jump to pad $N - 1$ with probability $\frac{N}{10}$ and to pad $N + 1$ with probability $1 - \frac{N}{10}$. Each jump is independent of the previous jumps. If the frog reaches pad 0 it will be eaten by a patiently waiting snake. If the frog reaches pad 10 it will exit the pond, never to return. What is the probability that the frog will escape without being eaten by the snake?

## Exercise 3.4.7

(AIME II 2019 Problem 2)

Lily pads 1, 2, 3, ... lie in a row on a pond. A frog makes a sequence of jumps starting on pad 1. From any pad $k$ the frog jumps to either pad $k + 1$ or pad $k + 2$ chosen randomly with probability $\frac{1}{2}$ and independently of other jumps. The probability that the frog visits pad 7 is $\frac{p}{q}$, where $p$ and $q$ are relatively prime positive integers. Find $p + q$.

## Exercise 3.4.8

(AIME II 2021 Problem 8)

An ant makes a sequence of moves on a cube where a move consists of walking from one vertex to an adjacent vertex along an edge of the cube. Initially the ant is at a vertex of the bottom face of the cube and chooses one of the three adjacent vertices to move to as its first move. For all moves after the first move, the ant does not return to its previous vertex, but chooses to move to one of the other two adjacent vertices. All choices are selected at random so that each of the possible moves is equally likely. The probability that after exactly 8 moves that ant is at a vertex of the top face on the cube is $\frac{m}{n}$, where $m$ and $n$ are relatively prime positive integers. Find $m + n$.

## Exercise 3.4.9

(HMMT February 2003 Combinatorics Problem 3)

Daniel and Scott are playing a game where a player wins as soon as he has two points more than his opponent. Both players start at par, and points are earned one at a time. If Daniel has a 60% chance of winning each point, what is the probability that he will win the game?

## Exercise 3.4.10

(USAMTS 3/4/20)

A particle is currently at the point $(0, 3.5)$ on the plane and is moving towards the origin. When the particle hits a lattice point (a point with integer coordinates), it turns with equal probability $45^\circ$ to the left or to the right from its current course. Find the probability that the particle reaches the $x$-axis before hitting the line $y = 6$.

## Example 3.5
(AMC 10A 2015 Problem 25)

Let S be a square of side length 1. Two points are chosen independently at random on the sides of S. The probability that the straight-line distance between the points is at least $\frac{1}{2}$ is $\frac{a-b\pi}{c}$, where $a$, $b$, and $c$ are positive integers and $\gcd(a,b,c)=1$. What is $a+b+c$?

(A) 59 (B) 60 (C) 61 (D) 62 (E) 63

## Discussion

**TIGER** The situation is very different depending on which sides the points are on, so we do casework on that. In some of the cases, we will use geometric probability. In any continuous probability problem (a problem in which something has a range of possible values instead of a countable set of them), geometric probability is almost always a good thing to start with.

**KEVIN** Here's a useful resource to learn geometry probability: https://brilliant.org/wiki/1-dimensional-geometric-probability/.

## Tiger's Solution

We have three cases.

**CASE 1:** The two points are on opposite sides. Then, the distance between them must be at least $\frac{1}{2}$.

**CASE 2:** The two points are on the same side. Choose a vertex of the square on the side containing both points. Then, let $x$ be the distance between the vertex and the first point, and let $y$ be the distance between the vertex and the second point. This is equivalent to choosing $x$ and $y$ randomly and uniformly between 0 and 1. The distance between the points is at least $\frac{1}{2}$ if and only if $|x-y| \geq \frac{1}{2}$. We can convert each ordered pair $(x,y)$ to a point on the coordinate plane. The possible values of $(x,y)$ are uniformly distributed in the unit square with coordinates between 0 and 1.

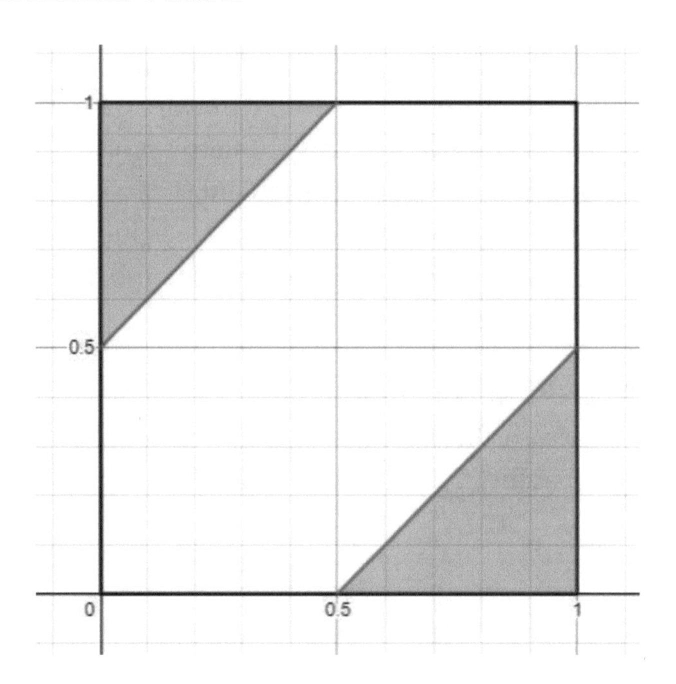


WWW.DESMOS.COM/CALCULATOR/QLTYOBKUEE

We want to find the area of the set of points on the graph of $|x-y| \geq \frac{1}{2}$ and in the square. This area is $2 \cdot \frac{1}{2} \cdot \frac{1}{2} \cdot \frac{1}{2} = \frac{1}{4}$. Thus, the probability that $|x-y| \geq \frac{1}{2}$ is $\frac{1}{4}$.

**CASE 3:** The two points are on adjacent sides. We can use geometric probability again. Choose the vertex of the square that is the intersection of the two sides that the vertices are on. Then, let $x$ be the distance between the vertex and the first point, and let $y$ be the distance between the vertex and the second point. This is equivalent to choosing $x$ and $y$ randomly and uniformly between 0 and 1. The distance between the points is at least $\frac{1}{2}$ if and only if $\sqrt{x^2+y^2} \geq \frac{1}{2}$, or $x^2+y^2 \geq \frac{1}{4}$. We can convert each ordered pair $(x,y)$ to a point on the coordinate plane. The possible values of $(x,y)$ are uniformly distributed in the unit square with coordinates between 0 and 1.

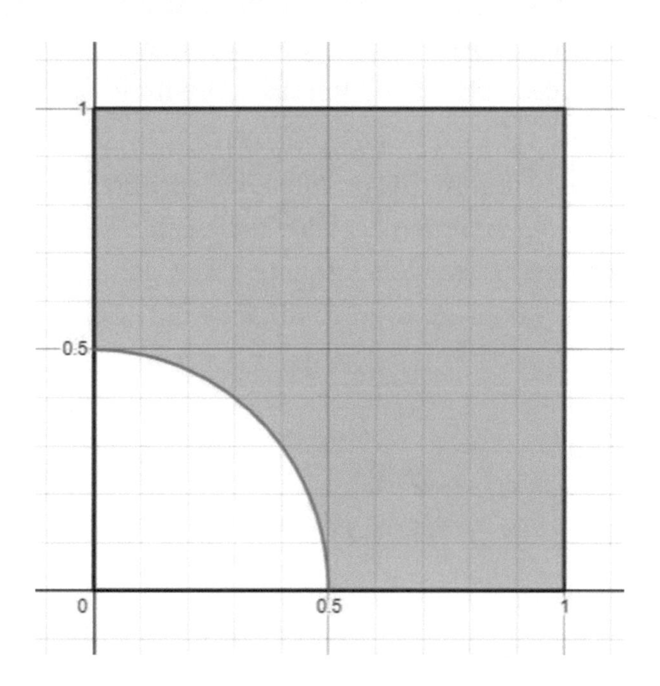


WWW.DESMOS.COM/CALCULATOR/SWNMCCYDT5

We want to find the area of the set of points on the graph of $x^2+y^2 \geq \frac{1}{4}$ and in the square. This area is $1-\frac{1}{4}\pi\left(\frac{1}{2}\right)^2 = 1-\frac{\pi}{16}$. Thus, the probability that $x^2+y^2 \geq \frac{1}{4}$ is $1-\frac{\pi}{16}$.

The probabilities that the two points are on opposite sides or on the same side are 1/4, and the probability that the two points are on adjacent sides is 1/2. Thus the probability that the two points are at least 1/2 units apart is:

$$\frac{1}{4}\cdot 1+\frac{1}{4}\cdot\frac{1}{4}+\frac{1}{2}\cdot\left(1-\frac{\pi}{16}\right)=\frac{26-\pi}{32},$$

giving an answer of (**A**) 59.

**KEVIN** When I tried this problem, I resolved case 3, the most interesting case in Tiger's solution, but I missed cases 1 and 2. Fortunately, my answer matched none of the choices, so I knew I missed something.

**TIGER** Notice that in this problem, we encoded the geometry into algebra then decoded the algebra back into geometry. We were left with a new geometry problem that was easier to solve than the old one.

## Exercises: Geometric Probability

### Exercise 3.5.1

(AMC 12 2001 Problem 17)

A point $P$ is selected at random from the interior of the pentagon with vertices $A=(0,2)$, $B=(4,0)$, $C=(2\pi+1,0)$, $D=(2\pi+1,4)$, and $E=(0,4)$. What is the probability that $\angle APB$ is obtuse?

## Exercise 3.5.2

(AMC 10B 2018 Problem 22)

Real numbers $x$ and $y$ are chosen independently and uniformly at random from the interval $[0, 1]$. Which of the following numbers is closest to the probability that $x$, $y$, and 1 are the side lengths of an obtuse triangle?

(A) 0.21 (B) 0.25 (C) 0.29 (D) 0.50 (E) 0.79

## Exercise 3.5.3

(Tiger)

Let $a$, $b$, and $c$ be randomly and independently chosen in the closed interval $[0, 1]$. What is the probability that they are the side lengths of an obtuse triangle?

## Exercise 3.5.4

(Tiger)

Let $a$ and $b$ be randomly and independently chosen in the closed interval $[0, 1]$. Given that $a^2 + b^2 \leq 1$, find the expected value of $\sqrt{1 - a^2 - b^2}$.

## Exercise 3.5.5

(AMC 10A 2019 Problem 22 & AMC 12A 2019 Problem 20)

Real numbers between 0 and 1, inclusive, are chosen in the following manner. A fair coin is flipped. If it lands heads, then it is flipped again and the chosen number is 0 if the second flip is heads, and 1 if the second flip is tails. On the other hand, if the first coin flip is tails, then the number is chosen uniformly at random from the closed interval $[0, 1]$. Two random numbers $x$ and $y$ are chosen independently in this manner. What is the probability that $|x - y| > \frac{1}{2}$?

## Exercise 3.5.6

(AMC 10B 2021 Problem 23)

A square with side length 8 is colored white except for 4 black isosceles right triangular regions with legs of length 2 in each corner of the square and a black diamond with side length $2\sqrt{2}$ in the center of the square, as shown in the diagram. A circular coin with diameter 1 is dropped onto the square and lands in a random location where the coin is completely contained within the square. The probability that the coin will cover part of the black region of the square can be written as $\frac{1}{196}\left(a + b\sqrt{2} + \pi\right)$, where $a$ and $b$ are positive integers. What is $a + b$?

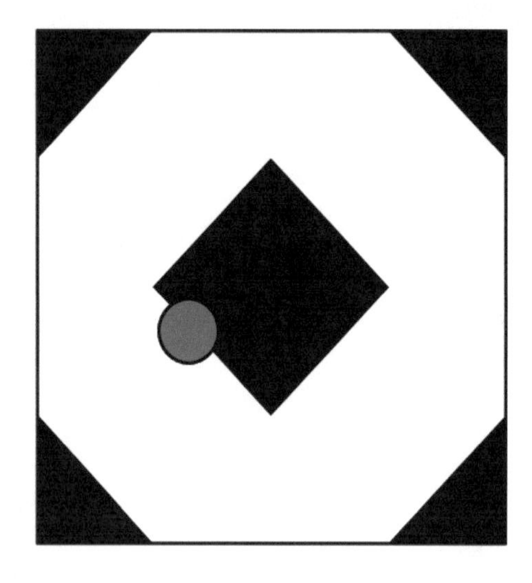

## Exercise 3.5.7

(AMC 10B 2022 Problem 23 & AMC 12B 2022 Problem 22)

Ant Amelia starts on the number line at 0 and crawls in the following manner. For $n = 1, 2, 3$, Amelia chooses a time duration $t_n$ and an increment $x_n$ independently and uniformly at random from the interval $(0, 1)$. During the $n$th step of the process, Amelia moves $x_n$ units in the positive direction, using up $t_n$ minutes. If the total elapsed time has exceeded 1 minute during the $n$th step, she stops at the end of that step; otherwise, she continues with the next step, taking at most 3 steps in all. What is the probability that Amelia's position when she stops will be greater than 1?

## Exercise 3.5.8

(AMC 12A 2016 Problem 23)

Three numbers in the interval $[0, 1]$ are chosen independently and at random. What is the probability that the chosen numbers are the side lengths of a triangle with positive area?

## Exercise 3.5.9

(AIME 1998 Problem 9)

Two mathematicians take a morning coffee break each day. They arrive at the cafeteria independently, at random times between 9 a.m. and 10 a.m., and stay for exactly $m$ minutes. The probability that either one arrives while the other is in the cafeteria is 40% and $m = a - b\sqrt{c}$ where $a$, $b$, and $c$ are positive integers, and $c$ is not divisible by the square of any prime. Find $a + b + c$.

## Exercise 3.5.10

(AIME I 2017 Problem 8)

Two real numbers $a$ and $b$ are chosen independently and uniformly at random from the interval $(0, 75)$. Let O and $P$ be two points on the plane with $OP = 200$. Let $Q$ and $R$ be on the same side of line $OP$ such that the degree measures of $\angle POQ$ and $\angle POR$ are $a$ and $b$ respectively, and $\angle OQP$ and $\angle ORP$ are both right angles. The probability that $QR \leq 100$ is equal to $\frac{m}{n}$, where $m$ and $n$ are relatively prime positive integers. Find $m + n$.

## Example 3.6

(AMC 10B 2019 Problem 25 & AMC 12B 2019 Problem 23)

How many sequences of 0s and 1s of length 19 are there that begin with a 0, end with a 0, contain no two consecutive 0s, and contain no three consecutive 1s?

(A) 55 (B) 60 (C) 65 (D) 70 (E) 75

## Discussion

**TIGER** Writing out cases for a 19-digit sequence seems difficult. (However, it actually isn't that difficult in this problem. Try it!) Instead, we use a technique called recursion, which will be demonstrated in the solution below.

**KEVIN** What made you think of trying recursion?

**TIGER** A lot of problems about sequences satisfying some conditions are solvable using recursion, but they must have the right conditions. For example, suppose we want to find the number of permutations of 1, 2, 3, 4, 5, 6, 7, 8, but with a condition. If the condition is that consecutive integers cannot be next to each other, then the problem is a good candidate for using recursion because the condition generalizes nicely to n-element permutations. If the condition says that 5 cannot be next to 6, then recursion won't work because the condition can't be nicely generalized.

## Tiger's Solution

We use recursion. Let $S_n$ be the number of sequences of length $n$ satisfying the properties stated in the problem. Consider a length $n$ sequence that satisfies these properties. We know that the second-last digit of the sequence must be 1. Let's take cases on what the third-last digit is.

**CASE 1:** The third-last digit is 0. We can ignore the last two numbers in the sequence and delete them. The resulting sequence (of length $n - 2$) begins and ends with a 0, so it must satisfy the properties stated in the problem. So, there are $S_{n-2}$ sequences that satisfy the properties.

**CASE 2:** The third-last digit is 1. The fourth-last digit must be 0, which means we can ignore the last three numbers in the sequence and delete them. The resulting sequence of length $n - 3$ begins and ends with a 0, so it must satisfy the properties stated in the problem. So, there are $S_{n-3}$ sequences that satisfy the properties.

Thus, we obtain the recurrence $S_n = S_{n-2} + S_{n-3}$. We know that $S_2 = 0, S_3 = 1$, and $S_4 = 1$, so we have

| $n$ | $S_n$ |
|---|---|
| 5 | 1 |
| 6 | 2 |
| 7 | 2 |
| 8 | 3 |
| 9 | 4 |
| 10 | 5 |
| 11 | 7 |
| 12 | 9 |
| 13 | 12 |
| 14 | 16 |
| 15 | 21 |
| 16 | 28 |
| 17 | 37 |
| 19 | 65 |

Thus, the answer is (C) 65.

**TIGER** Whenever a problem asks about a sequence with conditions of some sort, recursion might be helpful. It's even better if the sequence is of a fixed short length because that means iterating the recurrence fewer times. It's often possible to find closed-form formulas for recurrences, but that's mostly beyond the scope of the AMC.

# Exercises: Recursion

## Exercise 3.6.1

(AMC 8 2010 Problem 25)

Everyday at school, Jo climbs a flight of 6 stairs. Jo can take the stairs 1, 2, or 3 at a time. For example, Jo could climb 3, then 1, then 2. In how many ways can Jo climb the stairs?

## Exercise 3.6.2

(AMC 10A 2019 Problem 15 & AMC 12A 2019 Problem 9)

A sequence of numbers is defined recursively by $a_1 = 1$, $a_2 = \frac{3}{7}$, and $a_n = \frac{a_{n-2} \cdot a_{n-1}}{2a_{n-2} - a_{n-1}}$ for all $n \geq 3$. Then $a_{2019}$ can be written as $\frac{p}{q}$, where $p$ and $q$ are relatively prime positive integers. What is $p + q$?

## Exercise 3.6.3

(AMC 10B 2012 Problem 22)

Let $(a_1, a_2, ..., a_{10})$ be a list of the first 10 positive integers such that for each $2 \leq i \leq 10$ either $a_i + 1$ or $a_i - 1$ or both appear somewhere before $a_i$ in the list. How many such lists are there?

## Exercise 3.6.4

(Tiger)

Don writes the numbers 2, 3, ..., 9, in that order, on a blackboard. Don puts the operation $+$ or $\times$ between each two consecutive numbers, then calculates the result. In how many ways can Don put the operations to make the result even?

### Exercise 3.6.5

(Tiger)

Ten coins are arranged in a circle, all of them tails up. Every minute, Karen chooses a random coin that is tails up. Then, she looks at the coin and the two coins adjacent to it and flips any of these three coins that are tails up. What is the expected number of minutes that will pass until all the coins are heads up?

### Exercise 3.6.6

(AMC 12B 2021 Fall Problem 17)

A bug starts at a vertex of a grid made of equilateral triangles of side length 1. At each step the bug moves in one of the 6 possible directions along the grid lines randomly and independently with equal probability. What is the probability that after 5 moves the bug never will have been more than 1 unit away from the starting position?

### Exercise 3.6.7

(AMC 12A 2007 Problem 25)

Call a set of integers spacy if it contains no more than one out of any three consecutive integers. How many subsets of $\{1, 2, 3, \ldots, 12\}$, including the empty set, are spacy?

### Exercise 3.6.8

(AIME 1990 Problem 9)

A fair coin is to be tossed 10 times. Let $\frac{i}{j}$, in lowest terms, be the probability that heads never occur on consecutive tosses. Find $i + j$.

## Exercise 3.6.9

(AIME I 2006 Problem 11)

A collection of 8 cubes consists of one cube with edge-length $k$ for each integer $k$, $1 \leq k \leq 8$. A tower is to be built using all 8 cubes according to the rules:

- Any cube may be the bottom cube in the tower.
- The cube immediately on top of a cube with edge-length $k$ must have edge-length at most $k + 2$.

Let $T$ be the number of different towers than can be constructed. What is the remainder when $T$ is divided by 1000?

## Exercise 3.6.10

(AIME I 2016 Problem 13)

Freddy the frog is jumping around the coordinate plane searching for a river, which lies on the horizontal line $y = 24$. A fence is located at the horizontal line $y = 0$. On each jump Freddy randomly chooses a direction parallel to one of the coordinate axes and moves one unit in that direction. When he is at a point where $y = 0$, with equal likelihoods he chooses one of three directions where he either jumps parallel to the fence or jumps away from the fence, but he never chooses the direction that would have him cross over the fence to where $y < 0$.
Freddy starts his search at the point $(0, 21)$ and will stop once he reaches a point on the river. Find the expected number of jumps it will take Freddy to reach the river.

## Example 3.7
(AMC 10A 2012 Problem 25)

Real numbers $x$, $y$, and $z$ are chosen independently and at random from the interval $[0, n]$ for some positive integer $n$. The probability that no two of $x$, $y$, and $z$ are within 1 unit of each other is greater than $\frac{1}{2}$. What is the smallest possible value of $n$?

(A) 7 (B) 8 (C) 9 (D) 10 (E) 11

## Discussion

**TIGER** This problem is quite tricky, but its solution is clean and conceptual. First, we look at cases where $x \leq y \leq z$, and multiply by $3! = 6$ afterwards. We know that the distance between $x$ and $y$ is at least 1 and the distance between $y$ and $z$ is at least 1. That's quite annoying, so we want to remove this restriction. We can think of compressing the space between $x$ and $y$ until it is 1 unit less than before. Similarly, we can compress the space between $y$ and $z$ until it is 1 unit less than before. That is equivalent to translating the ordered triple $(x,y,z)$ to $(x',y',z') = (x,y-1,z-2)$. Then, we can use geometric probability intuition to finish.

## Tiger's Solution

First, we find the probability that $(x,y,z)$ satisfies the problem's conditions in terms of $n$. We look at the case $x \leq y \leq z$ only because all other cases are the same. Translate the ordered triple $(x,y,z)$ to

$(x', y', z') = (x, y-1, z-2)$. The conditions that $0 \leq x \leq y \leq z \leq n$, $y - x \geq 1$, and $z - y \geq 1$ can be replaced by the condition $0 \leq x' \leq y' \leq z' \leq n-2$.

Now, we can use the notion of geometric probability. Consider the following two solids:

1. The solid consisting of all possible points of $(x', y', z')$ with $x', y'$, and $z'$ in $[0, n-2]$ such that $x' \leq y' \leq z'$.
2. The solid consisting of all possible points of $(x, y, z)$ with $x, y$, and $z$ in $[0, n]$ such that $y - x \geq 1$, and $z - y \geq 1$.

Their volumes are equal because shifting all points on a figure in the same direction keeps its volume constant. The restriction $x' \leq y' \leq z'$ scales down the volume by a factor of $3! = 6$, so the volume of the solid containing all good points is $\frac{(n-2)^3}{6}$. Also, the volume of the space of possibilities of $(x, y, z)$ is $n^3$. Thus, the probability that $y - x \geq 1$, and $z - y \geq 1$ is :

$$\frac{\frac{(n-2)^3}{6}}{n^3}$$

Therefore, the probability that no two of $x, y$, and $z$ are within 1 unit of each other is:

$$6 \cdot \frac{\frac{(n-2)^3}{6}}{n^3} = \left(\frac{n-2}{n}\right)^3.$$

By inspection, the smallest positive integer $n$ satisfying $\left(\frac{n-2}{n}\right)^3 > \frac{1}{2}$ is (**D**) 10.

**KEVIN** Your approach of shifting variables is very nice. My first attempt used 3-D geometric probability with calculus.

**TIGER** The idea of shifting the variables can be motivated by stars and bars. In one stars and bars problem, we want to find the number of ways for $n$ indistinguishable balls to be placed into $k$ distinguishable boxes, and in another problem, there's an added restriction that each box has to have at least $m$ balls. However, the second problem can be reduced to the first one by placing $m$ balls in each of the $k$ boxes so that $n - mk$ balls remain. This is similar to the shifting idea I used on this problem.

**KEVIN** Stars and Bars is a counting problem that is useful by itself, but also has an instructive solution. See https: //en.wikipedia.org/wiki/Stars_and_bars_(combinatorics) and https: //artofproblemsolving.com/wiki/index.php/Ball-and-urn.

**TIGER** The idea of shifting is a specific case of a one-to-one correspondence—a map that converts one structure to another. By doing a transformation in this problem, we were also creating a one-to-one correspondence in disguise. There will be more one-to-one correspondences in the exercises. Here's one to keep in mind: any path from $(0, 0)$ to $(m, n)$ such that each step goes up one unit or to the right one unit can be converted to a string of $m$ R's and $n$ U's, where R represents a step to the right and each U represents a step up. Often, a one-to-one correspondence can be a simpler rephrasing of the problem.

# Exercises: One-to-One Correspondences

## Exercise 3.7.1
(AMC 8 2016 Problem 21)

A top hat contains 3 red chips and 2 green chips. Chips are drawn randomly, one at a time without replacement, until all 3 of the reds are drawn or until both green chips are drawn. What is the probability that the 3 reds are drawn?

## Exercise 3.7.2
(Tiger)

10 divers are lined up in 3 lines of 2, 3, and 5 people each. Every minute, one of the divers at the front of one of the lines jumps into the water. In how many orders can the divers jump into the water?

## Exercise 3.7.3
(AMC 10A 2016 Problem 20)

For some particular value of $N$, when $(a + b + c + d + 1)^N$ is expanded and like terms are combined, the resulting expression contains exactly 1001 terms that include all four variables $a, b, c$, and $d$, each to some positive power. What is $N$?

## Exercise 3.7.4

(AMC 10B 2020 Problem 23 & AMC 12B 2020 Problem 19)

Square $ABCD$ in the coordinate plane has vertices at the points $A(1,1)$, $B(-1,1)$, $C(-1,-1)$, and $D(1,-1)$. Consider the following four transformations:

- $L$, a rotation of $90^\circ$ counterclockwise around the origin;
- $R$, a rotation of $90^\circ$ clockwise around the origin;
- $H$, a reflection across the $x$-axis; and
- $V$, a reflection across the $y$-axis.

Each of these transformations maps the squares onto itself, but the positions of the labeled vertices will change. For example, applying $R$ and then $V$ would send the vertex $A$ at $(1,1)$ to $(-1,-1)$ and would send the vertex $B$ at $(-1,1)$ to itself. How many sequences of 20 transformations chosen from $\{L, R, H, V\}$ will send all of the labeled vertices back to their original positions? (For example, $R, R, V, H$ is one sequence of 4 transformations that will send the vertices back to their original positions.)

## Exercise 3.7.5

(AMC 10 2001 Problem 23 & AMC 12 2001 Problem 11)

A box contains exactly five chips, three red and two white. Chips are randomly removed one at a time without replacement until all the red chips are drawn or all the white chips are drawn. What is the probability that the last chip drawn is white?

## Exercise 3.7.6
(AMC 10B 2021 Problem 18)

A fair 6-sided die is repeatedly rolled until an odd number appears. What is the probability that every even number appears at least once before the first occurrence of an odd number?

## Exercise 3.7.7
(AIME I 2020 Problem 9)

Let $S$ be the set of positive integer divisors of $20^9$. Three numbers are chosen independently and at random with replacement from the set $S$ and labeled $a_1, a_2$, and $a_3$ in the order they are chosen. The probability that both $a_1$ divides $a_2$ and $a_2$ divides $a_3$ is $\frac{m}{n}$, where $m$ and $n$ are relatively prime positive integers. Find $m$.

## Exercise 3.7.8
(AIME 1984 Problem 11)

A gardener plants three maple trees, four oaks, and five birch trees in a row. He plants them in random order, each arrangement being equally likely. Let $\frac{m}{n}$ in lowest terms be the probability that no two birch trees are next to one another. Find $m + n$.

## Exercise 3.7.9
(Tiger)

Let $S$ be the set of all points with integer coordinates $(x, y, z)$ such that $0 \leq x \leq 8$, $0 \leq y \leq 1$, and $0 \leq z \leq 1$. Each point in $S$ is either shaded or not shaded. An "AoPS profile picture" is a shading such that if a point $(x, y, z)$ is shaded, then all other points $(x', y', z')$ in $S$ with $x' \leq x$, $y' \leq y$, and $z' \leq z$ is shaded. How many AoPS profile pictures are possible?

## Exercise 3.7.10

(HMMT February 2019 Combinatorics Problem 9)

How many ways can you fill a $3 \times 3$ square grid with nonnegative integers such that no nonzero integer appears more than once in the same row or column and the sum of the numbers in every row and column equals 7? (Note: this problem is very hard.)

*four*

# NUMBER THEORY

**KEVIN** Number theory is a big part of all levels of math competitions, including the AMC 10, though perhaps to a lesser extent than the other three subjects. Number theory deals with integers, the building blocks of mathematics. How do you feel about number theory problems?

**TIGER** I think the AMC and the AIME number theory problems have a more limited set of tricks we can apply, so computational number theory is easier to learn than a subject like geometry.

**KEVIN** Let's start with a problem involving digits.

## Example 4.1
(AMC 10B 2016 Problem 24)

How many four-digit integers *abcd*, with $a \neq 0$, have the property that the three two-digit integers $ab < bc < cd$ form an increasing arithmetic sequence? One such number is 4692, where $a = 4$, $b = 6$, $c = 9$, and $d = 2$.

(A) 9 (B) 15 (C) 16 (D) 17 (E) 20

## Discussion

**KEVIN** Problems involving digits and bases are usually solved by turning $a_n a_{n-1} \ldots a_0$ in base $b$ into $a_0 + b\, a_1 + b^2\, a_2 + \ldots + b^n\, a_n$ and using the fact that the digits are integers between 0 and $b - 1$, inclusive.

**TIGER** Once we get an equation, we usually want to find ways to reduce the number of cases to study.

**KEVIN** Tiger's method has fewer cases while my approach has more cases. Both work, but Tiger's method could be used as a standard approach for such problems.

## Tiger's Solution

Since $a \neq 0$ and $\underline{ab} < \underline{bc} < \underline{cd}$, we know that $1 \leq a \leq b \leq c \leq 9$. Since $\underline{ab}$, $\underline{bc}$, and $\underline{cd}$ form an arithmetic sequence, we have $2 \cdot \underline{bc} = \underline{ab} + \underline{cd}$, or $2(10b + c) = 10a + b + 10c + d$. We can bring all the multiples of 10 to one side to get $10(2b - a - c) = b - 2c + d$.

Since $0 \le b, c, d \le 9$, we have $-18 \le b - 2c + d \le 18$. The left-hand side is a multiple of 10, so $2b - a - c$ is $-1$, 0, or 1. Now, we've eliminated enough cases that it's feasible to test the remaining ones.

**CASE 1:** $2b - a - c = -1$ and $b - 2c + d = -10$. Since $b$ and $d$ are positive, we have $c \ge 6$. We check all possible values of $c$ to get the answers 2470, 1482, 3581, 5680, 2593, 4692, 6791, and 8890. This gives 8 possibilities.

**CASE 2:** $2b - a - c = 0$ and $b - 2c + d = 0$. We have $a + c = 2b$ and $b + d = 2c$, so $a$, $b$, $c$, and $d$ are an arithmetic progression. We also have $a \le b \le c$, so the arithmetic progression is nondecreasing. We do casework on the common difference of the arithmetic progression to get the answers 1234, 2345, 3456, 4567, 5678, 6789, 1357, 2468, and 3579. This gives 9 possibilities.

**CASE 3:** $2b - a - c = 1$ and $b - 2c + d = 10$. Since $b \le c$, we have $10 = b - 2c + d \le -c + d$, but $c$ and $d$ are decimal digits, a contradiction. Thus, there are no possibilities for Case 3.

In total, there are $8 + 9 =$ **(D)** 17 possibilities.

The key step in this problem is to factor the left-hand side, which restricts our possibilities a lot. A fundamental difference between digit problems and other number theory problems is that digits are bounded. Thus, we should exploit the boundedness condition and make a lot of reductions until there are few cases left.

## Kevin's Solution

Since $a \neq 0$ and $\underline{ab} < \underline{bc} < \underline{cd}$, we know $1 \le a \le b \le c \le 9$. Since $\underline{ab}$, $\underline{bc}$, and $\underline{cd}$ form an arithmetic sequence, we have $2 \cdot \underline{bc} = \underline{ab} + \underline{cd}$, or $2(10b + c) = 10a + b + 10c + d$. This rearranges to

$19b = 10a + 8c + d$. After a few minutes of bashing using casework on $b$, we get the answers $1234, 1357, 2345, 1482, 2468, 2470, 3456, 2593, 3579, 3581, 4567, 4692, 5678, 5680, 6789, 6791$, and $8890$, for a total of **(D)** 17 possibilities.

## Exercises: Digits and Bases

### Exercise 4.1.1
(Tiger)

Let $A$ and $B$ be digits such that both the sum and the positive difference of the two-digit numbers $\underline{AB}$ and $\underline{BA}$ are perfect squares. What is $A^2 + B^2$?

### Exercise 4.1.2
(AMC 10A 2011 Problem 23)

Seven students count from 1 to 1000 as follows:

- Alice says all the numbers, except she skips the middle number in each consecutive group of three numbers. That is, Alice says $1, 3, 4, 6, 7, 9, \ldots, 997, 999, 1000$.
- Barbara says all of the numbers that Alice doesn't say, except she also skips the middle number in each consecutive group of three numbers.
- Candice says all of the numbers that neither Alice nor Barbara says, except she also skips the middle number in each consecutive group of three numbers.

- Debbie, Eliza, and Fatima say all of the numbers that none of the students with the first names beginning before theirs in the alphabet say, except each also skips the middle number in each of her consecutive groups of three numbers.
- Finally, George says the only number that no one else says.

What number does George say?

## Exercise 4.1.3

(AMC 10A 2017 Problem 20)

Let $S(n)$ equal the sum of the digits of positive integer $n$. For example, $S(1507) = 13$. For a particular positive integer $n$, $S(n) = 1274$. Which of the following could be the value of $S(n + 1)$?

(A) 1 (B) 3 (C) 12 (D) 1239 (E) 1265

## Exercise 4.1.4

(Tiger)

Jessica is thinking of a number n. She gives the following clues:

- $n$ is less than $1000^{1000}$.
- When $n$ is written in base 1000, it contains only zeros and ones.
- When $n^2$ is written in base 1000, it contains only zeros, ones, and twos.

Let $N$ be the maximum value of $n$. How many ones are in the base 1000 representation of $N$?

## Exercise 4.1.5

(AMC 10B 2013 Problem 25)

Bernardo chooses a three-digit positive integer $N$ and writes both its base-5 and base-6 representations on a blackboard. Later LeRoy sees the two numbers Bernardo has written. Treating the two numbers as base-10 integers, he adds them to obtain an integer $S$. For example, if $N = 749$, Bernardo writes the numbers $10,444$ and $3,245$, and LeRoy obtains the sum $S = 13,689$. For how many choices of $N$ are the two rightmost digits of $S$, in order, the same as those of $2N$?

## Exercise 4.1.6

(Tiger)

How many nonnegative integers less than 1000 can be written as $100A + 10B + C$, where $A$, $B$, and $C$ are nonnegative integers that are in an arithmetic progression in that order? (The common difference does not have to be positive.)

## Exercise 4.1.7

(AMC 12B 2007 Problem 21)

The first 2007 positive integers are each written in base 3. How many of these base-3 representations are palindromes? (A palindrome is a number that reads the same forward and backward.)

## Exercise 4.1.8
(AIME I 2020 Problem 3)

A positive integer $N$ has base-eleven representation $\underline{abc}$ and base-eight representation $\underline{1bca}$ where $a, b$, and $c$ represent (not necessarily distinct) digits. Find the least such $N$ expressed in base ten.

## Exercise 4.1.9
(AIME I 2017 Problem 5)

A rational number written in base eight is $\underline{ab}.\underline{cd}$, where all digits are nonzero. The same number in base twelve is $\underline{bb}.\underline{ba}$ . Find the base-ten number $\underline{abc}$.

## Exercise 4.1.10
(AIME 1983 Problem 12)

Diameter $AB$ of a circle has length a 2-digit integer (base ten). Reversing the digits gives the length of the perpendicular chord $CD$. The distance from their intersection point $H$ to the center O is a positive rational number. Determine the length of $AB$.

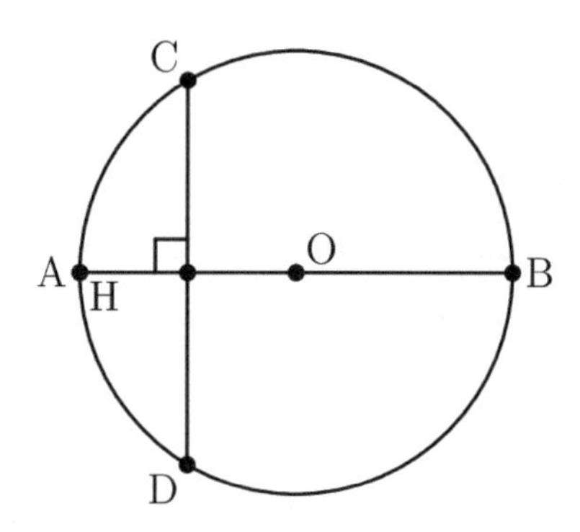

## Example 4.2

(AMC 10B 2017 Problem 25)

Last year Isabella took 7 math tests and received 7 different scores, each an integer between 91 and 100, inclusive. After each test she noticed that the average of her test scores was an integer. Her score on the seventh test was 95. What was her score on the sixth test?

(A) 92 (B) 94 (C) 96 (D) 98 (E) 100

## Discussion

**TIGER** Let $k$ be a weighted average of $x$ and $y$ with real number weights $\alpha$ and $\beta$ if $k = \alpha x + \beta y$ and $\alpha + \beta = 1$. I can introduce a way of thinking about weighted averages that will solve this problem easily. Its core is the following fact: if $k = \frac{z}{a+b}x + \frac{b}{z+b}y$, then $\frac{k-x}{y-k} = \frac{b}{a}$.

**KEVIN** Visually, we use a number line with two points of $X$, $Y$, and $K$ with coordinates $x, y$, and $k$, respectively, such that $K$ is in between $X$ and $Y$. If $k = \frac{a}{a+b}x + \frac{b}{a+b}y$, then the ratio of the distance from $X$ to $K$ to the distance from $Y$ to $K$ is $\frac{a}{b}$.

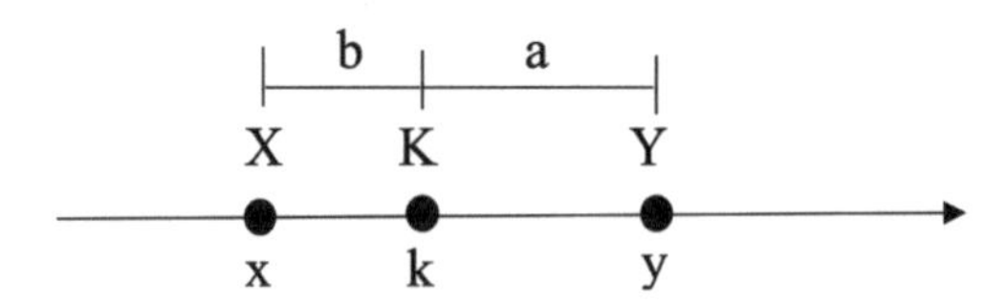


**TIGER** Let's do an example. In a list of seven numbers, the average of the first five is 29 and the average of the last two is 43. We want to find the average of all the numbers in the list. Let that average be $k$. Then, $k$ is a weighted average of 29 and 43 with a weight of $\frac{5}{7}$ on 29 and a weight of $\frac{2}{7}$ on 43. Thus, the distance between $k$ and 29, which is $k - 29$, is $\frac{2}{5}$ of the distance between $k$ and 43. Since the total distance between 29 and 43 is 14, we know that the distance between $k$ and 29 is $\frac{2}{7}$ of that. Thus, $k - 29 = \frac{2}{7} \cdot 14 = 4$, so $k = 33$.

We can think of this approach to weighted averages in the following way: suppose we have to find a weighted average of $x$ and $y$, where $x$ is weighted by $\frac{a}{a+b}$ and $y$ is weighted by $\frac{b}{a+b}$. Then, let $X$ and $Y$ be points on the number line that represent the numbers $x$ and $y$. Place a weight of $a$ on $X$ and a weight of $b$ on $Y$. The center of mass of the system (the point at which the weights at $X$ and $Y$ are balanced) is the weighted average of $x$ and $y$ that we were trying to find. This can be generalized to more dimensions and is the basis of barycentric coordinates, an algebraic method of solving geometry problems. (A diagram of a 2-D analog appears on the cover of this book.)

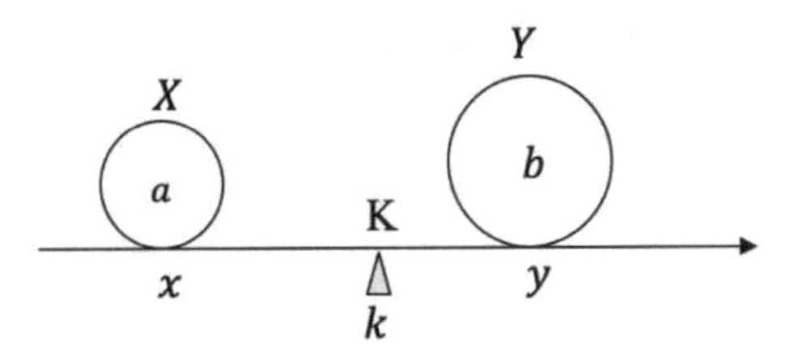


**KEVIN** First, I simplified the problem decreasing the scores by 90 to make them integers between 1 and 10, inclusive. Then, I used a more conventional method involving the Chinese remainder theorem:

> **CHINESE REMINDER THEOREM:**
>
> Let $n_1, n_2, \ldots, n_k$ be pairwise relatively prime positive integers. If $a_1, a_2, \ldots, a_k$ are integers, then the system
>
> $$\begin{aligned} x &\equiv a_1 \pmod{n_1}, \\ x &\equiv a_2 \pmod{n_2}, \\ &\vdots \\ x &\equiv a_k \pmod{n_k} \end{aligned}$$
>
> has a solution, and any two solutions are congruent modulo $n_1 n_2 \ldots n_k$.

Notice that the Chinese remainder theorem tells us that a solution to the system exists, but it doesn't tell us how to find it. This is usually done by trial and error.

## Tiger's Solution

We think about this problem backwards, starting from Isabella's seventh score. Let $y$ be average of her first 6 scores, and let $z$ be the average of her first 7 scores. We know that $y$ and $z$ are integers. We can rewrite $z$ as a weighted average of $y$ and her seventh score 95 with a $\frac{6}{7}$ weight on $y$ and a $\frac{1}{7}$ weight on 95. Thus, $z$ is $\frac{6}{7}$ of the way from 95 to $y$, so $y - 95$ must be divisible by 7. If $y \neq 95$, then it must be at least 102 or at most 88, which is impossible. Therefore $y = 95$.

Let $x$ be her sixth score. Then, we can write $y = 95$ as a weighted average of $x$ and the average of her first five scores $w$, with a weight of $\frac{5}{6}$ on $w$ and a weight of $\frac{1}{6}$ on $x$. Thus, $x - 95 = 5(95 - w)$. We know that $x$ is not 95 because the scores of the tests are distinct. The only other number from 91 to 100 that is a multiple of 5 away from 95 is **(E)** 100.

## Kevin's Solution

Let 7 integer scores be $a_1, a_2, a_3, a_4, a_5, a_6$, and $a_7$, and define $b_i = a_i - 90$, $i = 1, 2, \ldots, 6$. Then $1 \leq b_i \leq 10$, $i = 1, 2, \ldots, 6$, and $b_7 = 5$.

Since the average of $b_1, b_2, b_3, b_4, b_5, b_6$, and $b_7$ is an integer, we have $b_1 + b_2 + b_3 + b_4 + b_5 + b_6 + b_7 \equiv 0 \pmod 7$. Since $b_7 = 5$, we have $b_1 + b_2 + b_3 + b_4 + b_5 + b_6 \equiv 2 \pmod 7$.

Since the average of $b_1, b_2, b_3, b_4, b_5$, and $b_6$ is an integer, $b_1 + b_2 + b_3 + b_4 + b_5 + b_6 \equiv 0 \pmod 6$. By the Chinese remainder theorem $b_1 + b_2 + b_3 + b_4 + b_5 + b_6 \equiv 30 \pmod{42}$. Since $6 \leq b_1 + b_2 + b_3 + b_4 + b_5 + b_6 \leq 60$, we must have $b_1 + b_2 + b_3 + b_4 + b_5 + b_6 = 30$.

Since the average of $b_1, b_2, b_3, b_4$, and $b_5$ is an integer, $b_1 + b_2 + b_3 + b_4 + b_5 \equiv 0 \pmod 5$, so $b_6 = 30 - b_1 - b_2 - b_3 - b_4 - b_5 \equiv 0 \pmod 5$. We have $1 \leq b_6 \leq 10$ and $b_6 \neq b_7 = 5$, so we can only have $b_6 = 10$. Thus, $a_6 = 90 + b_6 =$ **(E)** 100.

# Exercises: Linear Number Theory

## Exercise 4.2.1

(Tiger)

Find $965 \cdot 3247 + 35 \cdot 4247$ using weighted averages.

## Exercise 4.2.2

(Tiger)

A sequence $a_1, a_2, \ldots$ satisfies $5a_{n+2} = 2a_{n+1} + 3a_n$ for all positive integers n. If $a_1 = 8$ and $a_4 = 65$, find the minimum possible value of $N$ such that for any positive odd integer n, $a_n \leq N$.

## Exercise 4.2.3

(AIME I 2017 Problem 9)

Let $a_{10} = 10$, and for each positive integer $n > 10$ let $a_n = 100a_{n-1} + n$. Find the least positive $n > 10$ such that $a_n$ is a multiple of 99.

## Exercise 4.2.4

(Tiger)

Alice and Bob are playing a game. Alice has one mango and Bob has 1002. Every minute, the person who has an even number of mangoes gives half of them to the other person. After how many minutes will Alice first have 1002 mangoes? .

## Exercise 4.2.5

(AIME I 2013 Problem 11)

Ms. Math's kindergarten class has 16 registered students. The classroom has a very large number, $N$, of play blocks which satisfies the conditions:

(A) If 16, 15, or 14 students are present in the class, then in each case all the blocks can be distributed in equal numbers to each student, and
(B) There are three integers $0 < x < y < z < 14$ such that when $x, y$, or $z$ students are present and the blocks are distributed in equal numbers to each student, there are exactly three blocks left over.

Find the sum of the distinct prime divisors of the least possible value of $N$ satisfying the above conditions.

## Example 4.3

(AMC 10B 2020 Problem 22)

What is the remainder when $2^{202} + 202$ is divided by $2^{101} + 2^{51} + 1$?

(A) 100 (B) 101 (C) 200 (D) 201 (E) 202

## Discussion

**TIGER** We can find the remainder by a strategy similar to Euclidean algorithm. The Euclidean algorithm is a process used to find the GCD of two numbers. It hinges on the fact that $\gcd(a, b) = \gcd(a, b - a)$. How do we prove this? We can do so by showing that the common divisors of $(a, b)$ and $(a, b - a)$ are the same, so their greatest common divisors are the same. The proof is short: if an integer $d$ divides $a$ and $b$, then it must divide $a$ and $a - b$; if $d$ divides $a$ and $a - b$, then it must divide $a$ and $a - (a - b) = b$.

**KEVIN** The Euclidean algorithm uses the above property to generate a series of steps to reduce two positive integers to smaller positive integers that have the same GCD. The steps continue until one of the integers is 0, with the other integer being the GCD.

Let's try to use this fact to find $\gcd(1961, 3811)$ without factoring. We have

$$\begin{aligned}\gcd(1961, 3811) &= \gcd(1961, 3811 - 1961)\\ &= \gcd(1961, 1850)\\ &= \gcd(1961 - 1850, 1850)\\ &= \gcd(111, 1850).\end{aligned}$$

We've simplified the expression enough to be able to find the GCD by factoring, but let's take the Euclidean algorithm a bit further for practice.

$$\begin{aligned}\gcd(111, 1850) &= \gcd(111, 1850 - 111)\\ &= \gcd(111, 1739)\\ &= \gcd(111, 1739 - 111)\\ &= \gcd(111, 1628).\end{aligned}$$

It's clear that we'll have to keep subtracting 111 for a long time. To speed this up, we can subtract multiples of 111.

$$\begin{aligned}\gcd(111, 1628) &= \gcd(111, 1628 - 14 \cdot 111)\\ &= \gcd(111, 74)\\ &= \gcd(111 - 74, 74)\\ &= \gcd(37, 74)\\ &= \gcd(37, 74 - 2 \cdot 37)\\ &= \gcd(37, 0).\end{aligned}$$

Once one of the numbers reaches 0, the algorithm stops. Now, it's clear that the largest number dividing both 37 and 0 is 37 (remember that 0 is divisible by every nonzero integer). Thus, $\gcd(1961, 3811) = 37$. The Euclidean algorithm is just this process.

**TIGER** We can generalize the Euclidean algorithm in many ways. It's useful for divisibility problems because of the following fact: $n$ is divisible by $d$ if and only if $\gcd(n, d) = d$. The Euclidean algorithm can also be extended to polynomials. That is, if we define what GCD is for polynomials. A polynomial $P(x)$ is divisible by a polynomial $D(x)$ if there exists a polynomial $Q(x)$ such that $P(x) = Q(x)D(x)$. We define $\gcd(P(x), Q(x))$ as a polynomial (usually, it's constrained to have a leading coefficient of 1) of highest degree that divides both $P(x)$ and $Q(x)$. Let's try to find $\gcd(x^2 + 3x - 4, x^3 - 2x + 1)$. We have

$$\begin{aligned}
&\gcd(x^2 + 3x - 4, x^3 - 2x + 1) \\
&\quad = \gcd(x^2 + 3x - 4, x^3 - 2x + 1 - x(x^2 + 3x - 4)) \\
&\quad = \gcd(x^2 + 3x - 4, -3x^2 + 2x + 1) \\
&\quad = \gcd(x^2 + 3x - 4, -3x^2 + 2x + 1 + 3(x^2 + 3x - 4)) \\
&\quad = \gcd(x^2 + 3x - 4, 11x - 11).
\end{aligned}$$

Notice that a polynomial divides $11x - 11$ if and only if it divides $x - 1$, so we have

$$\begin{aligned}
&\gcd(x^2 + 3x - 4, 11x - 11) = \gcd(x^2 + 3x - 4, x - 1) \\
&\quad = \gcd(x^2 + 3x - 4 - x(x - 1), x - 1) \\
&\quad = \gcd(4x - 4, x - 1) \\
&\quad = \gcd(4x - 4 - 4(x - 1), x - 1) \\
&\quad = \gcd(0, x - 1), \text{ which is just } x - 1.
\end{aligned}$$

### Tiger's Solution

We repeatedly subtract multiples of $2^{101}+2^{51}+1$:

$$(2^{202}+202)-2^{101}(2^{101}+2^{51}+1)=-2^{152}-2^{101}+202$$
$$(-2^{152}-2^{101}+202)+2^{51}(2^{101}+2^{51}+1)=2^{101}+2^{51}+202$$
$$(2^{101}+2^{51}+202)-(2^{101}+2^{51}+1)=201.$$

Therefore, the remainder is (D) 201.

## Exercises: Euclidean Algorithm

### Exercise 4.3.1

(Tiger)

Find $\gcd(x^{66}+x^{55}+x^{33}-x^{11}-1, x^{1381}+x^{1370}+x^{1348}+x^{1337})$.

### Exercise 4.3.2

(Jãnis Lazovskis)

Find all integer solutions to the equation $40x+25y=600$.

### Exercise 4.3.3

(IMO 1959 Problem 1)

Show that the fraction $\frac{21n+4}{14n+3}$ is irreducible for every positive integer $n$.

## Exercise 4.3.4

(AIME 1985 Problem 13)

The numbers in the sequence $101, 104, 109, 116, \ldots$ are of the form $a_n = 100 + n^2$, where $n = 1, 2, 3, \ldots$. For each $n$, let $d_n$ be the greatest common divisor of $a_n$ and $a_{n+1}$. Find the maximum value of $d_n$ as $n$ ranges through the positive integers.

## Exercise 4.3.5

Let $F_0 = 0$ and $F_1 = 1$, and let $F_n = F_{n-1} + F_{n-2}$ for all integers $n \leq 2$. Find gcd $\left(F_{40}, F_{45}\right)$.

## Exercise 4.3.6

(AIME I 2007 Problem 8)

The polynomial $P(x)$ is cubic. What is the largest value of $k$ for which the polynomials $Q_1(x) = x^2 + (k - 29)x - k$ and $Q_2(x) = 2x^2 + (2k - 43)x + k$ are both factors of $P(x)$?

## Exercise 4.3.7

(AIME II 2021 Problem 9)

Find the number of ordered pairs $(m, n)$ such that $m$ and $n$ are positive integers in the set $\{1, 2, \ldots, 30\}$ and the greatest common divisor of $2^m + 1$ and $2^n - 1$ is not 1.

## Exercise 4.3.8
(Tiger)

Let $n$ be a positive integer, and let $a_1, a_2, \ldots, a_n$ be a sequence such that for any integer $k$ with $1 \le k \le n$,

$$a_k = k(2n - 2k + 1).$$

Find the least 3-digit number $n$ such that any two consecutive numbers of the sequence are relatively prime.

## Exercise 4.3.9
(Tiger)

Let $\{a_n\}_{n>0}$ be a sequence of positive integers such that for any positive integers $i$ and $j$, $\gcd\left(a_i, a_{i+j}\right) \mid a_j$. Prove that for any positive integers $m$ and $n$, $\gcd\left(a_m, a_n\right) \mid a_{\gcd(m,n)}$.

## Exercise 4.3.10
(Inspired by IMO 2015 Problem 2, very hard)

Let $a$, $b$, and $c$ be positive integers such that $3 \le a \le b \le c$ and $bc - a$ is prime. Prove that $(ab - c)(ac - b)$ is not divisible by $bc - a$.

## Example 4.4
(AMC 10A 2018 Problem 22)

Let $a, b, c$, and $d$ be positive integers such that $\gcd(a, b) = 24$, $\gcd(b, c) = 36$, $\gcd(c, d) = 54$, and $70 < \gcd(d, a) < 100$. Which of the following must be a divisor of $a$?

(A) 5 (B) 7 (C) 11 (D) 13 (E) 17

## Discussion

**KEVIN** Here are some notation and theorems relating to the multiplicative structure of the positive integers.

For a prime $p$ and a nonzero integer $a$, define $\nu_p(a)$, known as the p-adic valuation of $a$, as the exponent of $p$ in the prime factorization of $a$. Then, for any prime $p$ and positive integers $a$ and $b$, we have $\nu_p(\gcd(a,b))$ $= \min(\nu_p(a), \nu_p(b))$ and $\nu_p(\text{lcm}(a,b)) = \max(\nu_p(a), \nu_p(b))$.

Notice that the infinite sequence $\nu_2(n)$, $\nu_3(n)$, $\nu_5(n)$, $\nu_7(n)$,... uniquely determines $n$. Thus, if we want to prove that two integers $a$ and $b$ are equal, it suffices to prove that $\nu_p(a) = \nu_p(b)$ for all primes $p$. The following facts can be proven from this:

1. $\gcd(ka, kb) = k \cdot \gcd(a, b)$ for all positive integers $a$, $b$, and $k$.
2. $\text{lcm}(ka, kb) = k \cdot \text{lcm}(a, b)$ for all positive integers $a$, $b$, and $k$.
3. $\gcd(a, b) \cdot \text{lcm}(a, b) = ab$ for all positive integers $a$ and $b$.

Also, try to prove that the number of positive divisors of $n$ is $(\nu_2(n) + 1)$ $(\nu_3(n) + 1)$ $(\nu_5(n) + 1)$... and the sum of all positive divisors of $n$ is $(1 + 2 + 2^2 + \ldots + 2^{\nu_2(n)})(1 + 3 + 3^2 + \ldots + 3^{\nu_3(n)})(1 + 5 + 5^2 + \ldots + 5^{\nu_5(n)})\ldots$.

**TIGER** Another common type of problem involves $\nu_p(n!)$. For example, if we want to find the number of trailing zeros of 100!, we can take the smaller number out of $\nu_2(100!)$ and $\nu_5(100!)$, which is $\nu_5(100!)$. Then, we can prove that $\nu_5(100!) = \left\lfloor \frac{100}{5} \right\rfloor + \left\lfloor \frac{100}{5^2} \right\rfloor + \left\lfloor \frac{100}{5^3} \right\rfloor + \ldots$, which gives the answer.

**KEVIN** My solution doesn't mention p-adic valuations but follows the same ideas.

## Tiger's Solution

Since $\gcd(a,b) = 24$ and $\gcd(b,c) = 36$, $a$ is divisible by 24 and $d$ is divisible by 54, so $a$ and $d$ are divisible by $\gcd(24, 54) = 6$ and $v_2(a), v_2(d), v_3(a), v_3(d) \geq 1$.

Since $\gcd(b,c) = 36$, $c$ is divisible by 36, so $v_2(c) \geq 2$. However, we also know that $v_2(54) = v_2(\gcd(\mathrm{c},\mathrm{d})) = \min(v_2(c), v_2(d)) = 1$, so $v_2(d) = 1$.

Similarly, since $\gcd(b,c) = 36$, $b$ is divisible by 36, so $v_3(b) \geq 2$. However, we also know that $v_3(24) = v_3(\gcd(a,b)) = \min(v_3(a), v_3(b)) = 1$, so $v_3(a) = 1$.

Since we have $v_2(d) = 1$ and $v_3(a) = 1$, we know that $v_2(\gcd(d,a)) = \min(v_2(d), v_2(a)) = 1$ and $v_3(\gcd(d,a)) = \min(v_3(d), v_3(a)) = 1$. The only value of $\gcd(d,a)$ between 70 and 100 satisfying this is 78. Thus, $a$ is divisible by 78, so $a$ must be divisible by **(D)** 13.

There are a lot of things we can do in this situation; the tricky part about some number theory problems is the number of things we can try. In this problem, the most important thing is to figure out the exact p-adic valuations of some variables, which will be useful since we can figure out nondivisibility, not just divisibility.

## Kevin's Solution

Notice that $24 = 2^3 \cdot 3$, $36 = 2^2 \cdot 3^2$, $54 = 2 \cdot 3^3$, and because $\gcd(a,b) = 24$, $\gcd(b,c) = 36$, and $\gcd(c,d) = 54$, $a$ is divisible by 24, $b$ is divisible by 72, $c$ is divisible by 108, and $d$ is divisible by 54, so we define $a = 2^3 \cdot 3 \cdot a_1$, $b = 2^3 \cdot 3^2 \cdot b_1$, $c = 2^2 \cdot 3^3 \cdot c_1$, $d = 2 \cdot 3^3 \cdot d_1$, where $a_1, b_1, c_1, d_1$ are positive integers that do not have prime factors 2 or 3.

Then,

$24 = \gcd(a,b) = \gcd(2^3 \cdot 3 \cdot a_1, 2^3 \cdot 3^2 \cdot b_1) = 2^3 \cdot 3 \cdot \gcd(a_1, 3b_1)$,
so $\gcd(a_1, 3b_1) = 1$;

$36 = \gcd(b,c) = \gcd(2^3 \cdot 3^2 \cdot b_1, 2^2 \cdot 3^3 \cdot c_1) = 2^2 \cdot 3^2 \cdot \gcd(2b_1, 3c_1)$,
so $\gcd(2b_1, 3c_1) = 1$;

$54 = \gcd(c,d) = \gcd(2^2 \cdot 3^3 \cdot c_1, 2 \cdot 3^3 \cdot d_1) = 2 \cdot 3^3 \cdot \gcd(2c_1, d_1)$,
so $\gcd(2c_1, d_1) = 1$.

So $\gcd(d,a) = \gcd(2 \cdot 3^3 \cdot d_1, 2^3 \cdot 3 \cdot a_1) = 2 \cdot 3 \cdot \gcd(9d_1, 4a_1)$. Neither $a_1$ nor $d_1$ has prime factors 2 and 3, so $\gcd(9d_1, 4a_1) = \gcd(d_1, a_1)$.
Because $70 < \gcd(d,a) < 100$, we must have $\frac{70}{6} < \gcd(d_1, a_1) < \frac{100}{6}$,
or $12 \le \gcd(d_1, a_1) \le 16$.

Among 12, 13, 14, 15, and 16, only 13 does not have any factor 2 or 3. So $\gcd(d_1, a_1) = 13$, **(D)** 13 is a factor of $a_1$, thus a factor of $a$.

## Exercises: Prime Factorization and p-adic Valuation

### Exercise 4.4.1

(AMC 8 2017 Problem 24)

Mrs. Sanders has three grandchildren, who call her regularly. One calls her every three days, one calls her every four days, and one calls her every five days. All three called her on December 31, 2016. On how many days during the next year did she not receive a phone call from any of her grandchildren?

## Exercise 4.4.2

(AMC 10A 2002 Problem 14 & AMC 12A 2002 Problem 12)

Both roots of the quadratic equation $x^2 - 63x + k = 0$ are prime numbers. The number of possible values of $k$ is

(A) 0 (B) 1 (C) 2 (D) 4 (E) more than 4

## Exercise 4.4.3

(AMC 10B 2019 Problem 19 & AMC 12B 2019 Problem 14)

Let $S$ be the set of all positive integer divisors of $100,000$. How many numbers are the product of two distinct elements of $S$?

## Exercise 4.4.4

(AMC 10B 2017 Problem 20 & AMC 12B 2017 Problem 16)

The number $21! = 51,090,942,171,709,440,000$ has over $60,000$ positive integer divisors. One of them is chosen at random. What is the probability that it is odd?

## Exercise 4.4.5

(AMC 10A 2016 Problem 22)

For some positive integer $n$, the number $110n^3$ has $110$ positive integer divisors, including $1$ and the number $110n^3$. How many positive integer divisors does the number $81n^4$ have?

## Exercise 4.4.6

(AMC 10B 2015 Problem 23)

Let $n$ be a positive integer greater than 4 such that the decimal representation of $n!$ ends in $k$ zeros and the decimal representation of $(2n)!$ ends in $3k$ zeros. Let $s$ denote the sum of the four least possible values of $n$. What is the sum of the digits of $s$?

## Exercise 4.4.7

(AMC 10A 2019 Problem 25 & AMC 12A 2019 Problem 24)

For how many integers $n$ between 1 and 50, inclusive, is $\frac{(n^2-1)!}{(n!)^n}$ an integer? (Recall that $0! = 1$.)

## Exercise 4.4.8

(AIME II 2004 Problem 8)

How many positive integer divisors of $2004^{2004}$ are divisible by exactly 2004 positive integers?

## Exercise 4.4.9

(AIME I 2021 Problem 14)

For any positive integer $a$, $\sigma(a)$ denotes the sum of the positive integer divisors of $a$. Let $n$ be the least positive integer such that $\sigma(a^n) - 1$ is divisible by 2021 for all positive integers $a$. Find the sum of the prime factors in the prime factorization of $n$.

## Exercise 4.4.10
(IMO 1990 Problem 3)

Determine all integers $n > 1$ such that $\frac{2^n + 1}{n^2}$ is an integer.

## Example 4.5
(AMC 10B 2018 Problem 21)

Mary chose an even 4-digit number $n$. She wrote down all the divisors of $n$ in increasing order from left to right: $1, 2, \ldots, \frac{n}{2}, n$. At some moment Mary wrote 323 as a divisor of $n$. What is the smallest possible value of the next divisor written to the right of 323?

(A) 324 (B) 330 (C) 340 (D) 361 (E) 646

## Discussion

**KEVIN** How quickly can you find the divisors of 323? Doing so is required for this problem. I ask my students to remember the square numbers less than 1000. If they remember that $324 = 18^2$, then they can use the difference of squares formula to get

$$323 = 324 - 1 = 18^2 - 1^2 = (18 - 1)(18 + 1) = 17 \cdot 19.$$

**TIGER** Suppose the next divisor after 323 is $k$. We know that $n$ is divisible by both 323 and $k$, so $n$ is divisible by $\operatorname{lcm}(323, k)$. Later in my solution, I also involved $\gcd(323, k)$ using $\gcd(a,b) \cdot \operatorname{lcm}(a,b) = ab$.

**KEVIN** My solution did not involve LCM or GCD, but used factoring and estimation based on the prime factorization of $n$.

## Tiger's Solution

Suppose the next divisor after 323 is $k$. Then, we know that $n$ is divisible by both 323 and $k$, so $n$ is divisible by $\text{lcm}(323, k)$. Since $n < 10000$, we want to choose $k$ that would make $\text{lcm}(323, k)$ small. Intuitively, that would mean 323 and $k$ have many shared factors. We can formalize this by bringing GCD into the expression.

Since $\gcd(a,b) \cdot \text{lcm}(a,b) = ab$, we have $\text{lcm}(323,k) = \frac{323k}{\gcd(323,k)}$. Since $k > 323$, we know that $323k > 323 \cdot 323 > 10000$. Since $\text{lcm}(323, k) < 10000$, we know that $\gcd(323, k) > 10$.

Since $323 = 17 \cdot 19$, we know that $\gcd(323, k)$ is either 17, 19, or 323. If $\gcd(323, k) = 17$, then $k = 323 + 17 = 340$ is the least possible value. If $\gcd(323, k) = 19$, then $k = 323 + 19 = 342$ is the least possible. If the $\gcd(323, k) = 323$, then $323 + 323 = 646$ is the least possible. If $k = 340$, then $n = 6460 = 2 \cdot 2 \cdot 5 \cdot 17 \cdot 19$. Then, 323 and 340 are factors of 6460, so $k =$ (C) 340 can be achieved.

## Kevin's Solution

Since $n$ is an even multiple of 323, we can write $n = 646n_1$ for a positive integer $n_1$. Since $1000 \leq n \leq 9999$, $\frac{1000}{646} \leq n_1 \leq \frac{9999}{646}$, or $2 \leq n_1 \leq 15$.

For the smallest divisor of $n$ to be larger than $17 \cdot 19$, we can try $17 \cdot 20$ with $n_1 = 10$ or $18 \cdot 19$ with $n_1 = 9$. First one is smaller with $n = 6460$. So the next divisor to the right of 323 is $17 \cdot 20 =$ (C) 340.

## Exercises: GCD and LCM

### Exercise 4.5.1
(AMC 8 2016 Problem 20)

The least common multiple of $a$ and $b$ is 12, and the least common multiple of $b$ and $c$ is 15. What is the least possible value of the least common multiple of $a$ and $c$?

### Exercise 4.5.2
(JMPSC Invitationals 2021 Problem 3, by Tiger)

There are exactly 5 even positive integers less than or equal to 100 that are divisible by $x$. What is the sum of all possible positive integer values of $x$?

### Exercise 4.5.3
(Tiger)

What is the sum of all nonnegative integers c such that $x^2 - 85x + c^2$ has at least one integer root?

### Exercise 4.5.4
(hackmath.net)

The least common multiple of two numbers is 22 more than their greatest common divisor. How many pairs of such two numbers are there?

### Exercise 4.5.5

(AMC 12A 2020 Problem 21)

How many positive integers $n$ are there such that $n$ is a multiple of 5, and the least common multiple of $5!$ and $n$ equals 5 times the greatest common divisor of $10!$ and $n$?

### Exercise 4.5.6

(AMC 10B 2018 Problem 23)

How many ordered pairs $(a, b)$ of positive integers satisfy the equation $a \cdot b + 63 = 20 \cdot \text{lcm}\,(a, b) + 12 \cdot \gcd\,(a, b)$, where $\gcd\,(a, b)$ denotes the greatest common divisor of $a$ and $b$, and $\text{lcm}\,(a, b)$ denotes their least common multiple?

### Exercise 4.5.7

(AMC 10A 2016 Problem 25)

How many ordered triples $(x, y, z)$ of positive integers satisfy $\text{lcm}(x, y) = 72$, $\text{lcm}(x, z) = 600$, and $\text{lcm}(y, z) = 900$?

### Exercise 4.5.8

(AIME 1986 Problem 5)

What is the largest positive integer $n$ for which $n^3 + 100$ is divisible by $n + 10$?

## Exercise 4.5.9

(AIME 1987 Problem 7)

Let $[r, s]$ denote the least common multiple of positive integers $r$ and $s$. Find the number of ordered triples $(a, b, c)$ of positive integers for which $[a, b] = 1000$, $[b, c] = 2000$, and $[c, a] = 2000$.

## Exercise 4.5.10

(AIME I 2020 Problem 10)

Let $m$ and $n$ be positive integers satisfying the conditions

- $\gcd(m + n, 210) = 1$,
- $m^m$ is a multiple of $n^n$, and
- $m$ is not a multiple of $n$.

Find the least possible value of $m + n$.

## Example 4.6

(AMC 10A 2012 Problem 22)

The sum of the first $m$ positive odd integers is 212 more than the sum of the first $n$ positive even integers. What is the sum of all possible values of $n$?

(A) 255 (B) 256 (C) 257 (D) 258 (E) 259

## Discussion

**TIGER** We start by setting up an equation that we want to find integer solutions to. This is called a Diophantine equation.

**KEVIN** In Chapter 2, we introduced Simon's favorite factoring trick to solve Diophantine equations. I followed the same idea and factored the equation by completing the square and using difference of squares.

**TIGER** I used the quadratic formula to handle the second-degree Diophantine equation, which is also a common method.

## Tiger's Solution

The sum of the first $m$ positive odd integers is $m^2$, and the sum of the first $n$ positive even integers is $n^2 + n$. Thus, we have $m^2 = n^2 + n + 212$. Consider this equation as a quadratic in $n$. Since $n$ is an integer, the discriminant $1^2 - 4(212 - m^2) = 4m^2 - 847$ must be the square of an integer.

Let $k$ be the nonnegative integer with $4m^2 - 847 = k^2$. Then, we have $847 = 4m^2 - k^2 = (2m - k)(2m + k)$. Since $847 = 7 \cdot 11^2$, the only pairs that multiply to 847 are $(1, 847)$, $(7, 121)$, and $(11, 77)$. Thus, the possible values of $(m, k)$ are $(212, 423)$, $(32, 57)$, and $(22, 33)$.

We solve the original Diophantine equation using the quadratic formula to get $n = \frac{-1 \pm \sqrt{4m^2 - 847}}{2} = \frac{-1 \pm k}{2}$. Therefore, the possible values of $n$ are 211, 28, and 16, which sum up to (A) 255.

**TIGER** Here are some tips for solving Diophantine equations:

1. It's often useful to write a variable in terms of the other variables, especially in a Diophantine equation with two variables. In this problem, we used the quadratic formula to get $n$ in terms of $m$.
2. Find a way to factor (often using difference of $n$th powers or Simon's favorite factoring trick). It's especially helpful if the right-hand side is a constant. In this problem, we found that $2m - k$ and $2m + k$ multiply to a constant, 847.
3. Take the whole expression modulo some number; that number can even be a variable. This strategy is sometimes used when proving that a Diophantine equation has no solutions or very few solutions.
4. Sometimes, number theory might not be particularly helpful in the beginning. Instead, use algebra to simplify the equation to a form in which number theory will be helpful.

## Kevin's Solution

The sum of the first $m$ positive odd integers is $m^2$, and the sum of the first $n$ positive even integers is $n^2 + n$. Thus, we have $m^2 = n^2 + n + 212 \Rightarrow m^2 - n^2 - n = 212$, so

$$\frac{847}{4} = m^2 - n^2 - n - \frac{1}{4} = m^2 - \left(n + \frac{1}{2}\right)^2 = \left(m + n + \frac{1}{2}\right)\left(m - n - \frac{1}{2}\right).$$

We multiply by 4 on both sides to get $(2m + 2n + 1)(2m - 2n - 1) = 847$. Since $847 = 7 \cdot 11^2$, the only pairs that multiply to 847 are $(1, 847)$,

$(7, 121)$, and $(11, 77)$. Thus, we have the solutions $(m, n) = (212, 211)$, $(32, 28)$, and $(22, 16)$, so the sum of all possible $n$ is $211 + 28 + 16 =$ (**A**) 255.

## Exercises: Diophantine Equations

### Exercise 4.6.1

(AMC 8 2019 Problem 23)

After Euclid High School's last basketball game, it was determined that $\frac{1}{4}$ of the team's points were scored by Alexa and $\frac{2}{7}$ were scored by Brittany. Chelsea scored 15 points. None of the other 7 team members scored more than 2 points. What was the total number of points scored by the other 7 team members?

### Exercise 4.6.2

(AMC 10B 2013 Problem 21 & AMC 12B 2013 Problem 14)

Two non-decreasing sequences of nonnegative integers have different first terms. Each sequence has the property that each term beginning with the third is the sum of the previous two terms, and the seventh term of each sequence is $N$. What is the smallest possible value of $N$?

### Exercise 4.6.3

(AMC 10B 2022 Problem 7 & AMC 12B 2022 Problem 4)

For how many values of the constant $k$ will the polynomial $x^2 + kx + 36$ have two distinct integer roots?

## Exercise 4.6.4

(AMC 10A 2021 Problem 22)

Hiram's algebra notes are 50 pages long and are printed on 25 sheets of paper; the first sheet contains pages 1 and 2, the second sheet contains pages 3 and 4, and so on. One day he leaves his notes on the table before leaving for lunch, and his roommate decides to borrow some pages from the middle of the notes. When Hiram comes back, he discovers that his roommate has taken a consecutive set of sheets from the notes and that the average (mean) of the page numbers on all remaining sheets is exactly 19. How many sheets were borrowed?

## Exercise 4.6.5

(AIME II 2000 Problem 2)

A point whose coordinates are both integers is called a lattice point. How many lattice points lie on the hyperbola $x^2 - y^2 = 2000^2$?

## Exercise 4.6.6

(AMC 10A 2015 Problem 24 & AMC 12A 2015 Problem 19)

For some positive integers $p$, there is a quadrilateral $ABCD$ with positive integer side lengths, perimeter $p$, right angles at $B$ and $C$, $AB = 2$, and $CD = AD$. How many different values of $p < 2015$ are possible?

## Exercise 4.6.7

(AMC 10B 2007 Problem 25)

How many pairs of positive integers $(a, b)$ are there such that $a$ and $b$ have no common factors greater than 1 and:

$$\frac{a}{b} + \frac{14b}{9a}$$

is an integer?

## Exercise 4.6.8

(HMMT February 2018 Algebra and Number Theory Problem 4)

Distinct prime numbers $p$, $q$, $r$ satisfy the equation

$$2pqr + 50pq = 7pqr + 55pr = 8pqr + 12qr = A$$

for some positive integer $A$. What is $A$?

## Exercise 4.6.9

(Tiger)

Let $a$, $b$, $c$, and $d$ be integers such that $a^3 + b^3 = 7c + 7d \neq 0$ and $c^3 + d^3 = 7a + 7b \neq 0$. Find, with proof, all possible values of $a + b + c + d$.

## Exercise 4.6.10

(AIME I 2011 Problem 15)

For some integer $m$, the polynomial $x^3 - 2011x + m$ has the three integer roots a, $b$, and $c$. Find $|a| + |b| + |c|$.

# CLOSING

**KEVIN** Well, we planned to finish this book during the summer of 2021, and it has taken a lot longer. During this extended period of time, I have taught hundreds more students and you have achieved more in math competitions.

**TIGER** I really enjoyed writing this book with you!

**KEVIN** What's next for you?

**TIGER** I enjoy my high school life, which has been quite busy. I study Olympiad math, do math research, practice piano, and participate in performances. How about you?

**KEVIN** I am currently teaching Algebra 1 and two new courses, AP Statistics and Financial Algebra. Teaching new subjects is always challenging. Next school year, I plan to be an academic competition coach at my new school, Calabasas High School. Plus, I always have a number of students who I tutor after school. Also, I'll publish this book via a number of channels. Do you anticipate any challenges in the coming months?

**TIGER** There will always be challenges in math or music or any other high school activities, but I'm not afraid of them. If I fail in any such challenges, I feel sad for a moment, but then I move on and start thinking about how to improve next time.

**KEVIN** For a 10th grader, your achievements in math have been outstanding. Can you share how you learned math?

**TIGER** I mostly learned math through self-study. I also got some early exposure from the UCLA Math Circle and benefited from various AoPS courses.

**KEVIN** When your mom brought you to my first AIME class when I was teaching at the UCLA Math Circle, you were only in 5th grade even though the AIME is usually for high school students. But she wanted you to try the class anyway. I said that if you could solve one of the problems, you could stay, and you did indeed solve one. Do you remember that problem (see Exercise 5.1.1)?

**TIGER** No, but I do remember that in the first homework, I solved an AIME problem 11 (see Exercise 5.1.4).

**KEVIN** Back then, did you feel that the AIME was hard?

**TIGER** Yes, it was hard in the beginning. But if I keep trying to solve hard problems, those I previously thought were almost impossible became manageable.

**KEVIN** You qualified for the AIME the very next year when you were in 6th grade. The AIME is a three-hour exam and many of my students can't concentrate on math problems for that long. How are you able to focus on math problems for a long time without getting tired?

**TIGER** The most important thing is to enjoy math; if you do, then you can work for a long time. It can be helpful and fun to practice by looking at a problem that's way above your level (like an Olympiad problem) and try to solve it. Just look at it every now and then, and don't be discouraged if you can't make any progress for a long time. I saw someone on Numberphile, a YouTube math channel, who took a whole year to solve the last problem of IMO 1988. I need to add that you should work on these problems because they are fun, not because you want to practice being patient. That's the only way it will work well for you.

However, not all Olympiad problems are as approachable, and some require advanced techniques. Some good problems to try (whose techniques don't exceed the AMC 10 level) are listed in the Exercises.

**KEVIN** I could imagine that playing in piano competitions is a high-pressure situation. You also told me that you had only 3-4 minutes to do problem 25 in the 2021 the AMC 10B (see Example 1.6). That was a hard problem and many students might panic and be unable to think straight under the time pressure, but you were able to solve it. Do you think your experience in piano competitions helped you in math competitions?

**TIGER** It might have helped, but I think what helped me most is that I don't require myself to do all 25 problems correctly. I just tried to do my best.

**KEVIN** Congratulations on becoming the only one in the world who scored perfectly in the AMC 10B Spring 2021!

**TIGER** Thank you!

## Exercises: Tiger's First AIME Practice Problems

### Exercise 5.1.1

(AIME I 2011 Problem 1)

Jar A contains four liters of a solution that is 45% acid; jar B contains five liters of a solution that is 48% acid; jar C contains one liter of a solution that is $k\%$ acid. From jar C, $\frac{m}{n}$ liters of the solution is added to jar A, and the remainder of the solution in jar C is added to jar B. At the end, both jar A and jar B contain solutions that are 50% acid. Given that $m$ and $n$ are relatively prime positive integers, find $k + m + n$.

### Exercise 5.1.2

(AIME I 2011 Problem 2)

In rectangle $ABCD$, $AB = 12$ and $BC = 10$. Points $E$ and $F$ lie inside rectangle $ABCD$ so that $BE = 9$, $DF = 8$, $\overline{BE} \parallel \overline{DF}$, $\overline{EF} \parallel \overline{AB}$, and line $BE$ intersects segment $\overline{AD}$. The length $EF$ can be expressed in the form

$m\sqrt{n} - p$, where $m, n$, and $p$ are positive integers and $n$ is not divisible by the square of any prime. Find $m + n + p$.

### Exercise 5.1.3

(AIME I 2011 Problem 3)

Let $L$ be the line with slope $\frac{5}{12}$ that contains the point $A = (24, -1)$, and let $M$ be the line perpendicular to line $L$ that contains the point $B = (5, 6)$. The original coordinate axes are erased, and line $L$ is made the $x$-axis, and line $M$ the $y$-axis. In the new coordinate system, point $A$ is on the positive $x$-axis, and point $B$ is on the positive $y$-axis. The point $P$ with coordinates $(-14, 27)$ in the original system has coordinates $(\alpha, \beta)$ in the new coordinate system. Find $\alpha + \beta$.

### Exercise 5.1.4

(AIME I 2011 Problem 11)

Let $\mathscr{R}$ be the set of all possible remainders when a number of the form $2^n$, $n$ a nonnegative integer, is divided by 1000. Let $\mathscr{S}$ be the sum of the elements in $\mathscr{R}$. Find the remainder when $\mathscr{S}$ is divided by 1000.

## Exercises: Fun Problems to Think About for a Loooong Time

### Exercise 5.2.1

(USAJMO 2021 Problem 4)

Carina has three pins, labeled $A, B$, and $C$, respectively, located at the origin of the coordinate plane. In a *move*, Carina may move a pin to an adjacent lattice point at distance 1 away. What is the least number of moves that Carina can make in order for triangle $ABC$ to have area 2021?

(A lattice point is a point $(x, y)$ in the coordinate plane where $x$ and $y$ are both integers, not necessarily positive.)

## Exercise 5.2.2

(IMO 2017 Problem 1)

For each integer $a_0 > 1$, define the sequence $a_0, a_1, a_2, \cdots$ for $n \geq 0$ as

$$a_{n+1} = \begin{cases} \sqrt{a_n} & \text{if } \sqrt{a_n} \text{ is an integer,} \\ a_n + 3 & \text{otherwise.} \end{cases}$$

Determine all values of $a_0$ such that there exists a number $A$ such that $a_n = A$ for infinitely many values of $n$.

## Exercise 5.2.3

(IMO 2018 Problem 1)

Let $\Gamma$ be the circumcircle of acute triangle $ABC$. Points $D$ and $E$ are on segments $AB$ and $AC$ respectively such that $AD = AE$. The perpendicular bisectors of $BD$ and $CE$ intersect minor arcs $AB$ and $AC$ of $\Gamma$ at points $F$ and $G$ respectively. Prove that lines $DE$ and $FG$ are either parallel or they are the same line.

## Exercise 5.2.4
(IMO Shortlist 2018 A4)

Let $a_0, a_1, a_2, \cdots$ be a sequence of real numbers such that $a_0 = 0, a_1 = 1$ and for every $n \geq 2$ there exists $1 \leq k \leq n$ satisfying

$$a_n = \frac{a_{n-1} + \cdots + a_{n-k}}{k}.$$

Find the maximum possible value of $a_{2018} - a_{2017}$.

## Exercise 5.2.5
(USA TSTST 2017 Problem 6)

A sequence of positive integers $(a_n)_{n\geq 1}$ is of *Fibonacci type* if it satisfies the recursive relation $a_{n+2} = a_{n+1} + a_n$ for all $n \geq 1$. Is it possible to partition the set of positive integers into an infinite number of Fibonacci type sequences?